COMMENTS ON *GREAT VIDEOS FOR KIDS*

"Well written, in an easy to read, concise format, Catherine Cella's book, *Great Videos for Kids,* is an invaluable resource . . . Anyone concerned about what children are watching on videos should own this book!"

—Paula Miller, executive director
Coalition for Quality Children's Videos

"Catherine Cella's book is an excellent resource . . . It successfully guides anyone interested in children's programs toward the best and appropriate kid videos. Reading her book makes me proud to be involved in this industry, and I thank Catherine for compiling this accurate and expert resource."

—Amy Weintraub, president, Backyard Productions
producers of "Baby Songs" and "Tales & Tunes"

"What an incredible source for parents and teachers seeking quality videos for children. Catherine Cella is very insightful—she sees what children are learning beyond the storyline. *Great Videos for Kids* truly is a gem."

—Susan Nipp, Pam Beall
"Wee Sing" authors

"What a great time-saving tool . . . Now you don't have to pore through video bins searching for something appropriate for your kids. Cella's done the legwork!"

—Rona Elliot, entertainment reproter
ABC/Home Show

"For all those who care that a child's television experience should be engaging and enriching, as well as entertaining, *Great Videos for Kids* is a wonderful instruction manual. By including children in the formula for selecting videos, Cella provides reassurance that the videos you choose won't simply sit on a shelf. (*Great Videos for Kids*) is the best book on the subject."

—Vivian M. Horner, Ph.D, creator of Nickelodeon
senior vice president, Development, SkyPix

"This comprehensive compendium is a MUST-HAVE, MUST-USE guide for anyone who cares for and about what our children are watching on television these days. Keep a copy handy in the car en route to the library or the video store!"

—Morton Schindel, founder
Children's Circle and Weston Woods

"Catherine Cella has provided parents, and all people who work with young children, a wonderful, quick, easy-to-use comprehensive reference for selecting vidcos for young children. It is exciting to know that all of this information is contained in one compact book!"

—Kathy Parker, executive producer
"Barney & Friends"

"*Great Videos for Kids* is a concise and much-needed guide through the largely uncharted world of children's video. A terrific help for all parents."

—Steven H. Scheuer, television critic and
author of *Movies on TV and Videocassette*

Great Videos for Kids

GREAT VIDEOS FOR KIDS

A Parent's Guide to Choosing the Best

by Catherine Cella

A Citadel Press Book
Published by Carol Publishing Group

A Citadel Press Book
Published by Carol Publishing Group
Citadel Press is a registered trademark of Carol Communications, Inc.
Editorial Offices: 600 Madison Avenue, New York, N.Y. 10022
Sales and Distribution Offices: 120 Enterprise Avenue, Secaucus, N.J. 07094
In Canada: Canadian Manda Group, P.O. Box 920, Station U, Toronto, Ontario
M8Z 5P9
Queries regarding rights and permissions should be addressed to Carol Publishing Group, 600 Madison Avenue, New York, N.Y. 10022

Carol Publishing Group books are available at special discounts for bulk purchases, for sales promotions, fund raising, or educational purposes. Special editions can be created to specifications. For details, contact Special Sales Department, Carol Publishing Group, 120 Enterprise Avenue, Secaucus, N.J. 07094

Some of the reviews in this book first appeared in the following publications: *Billboard* magazine, *Parents' Guide to Children's Entertainment* (800-966-5209), *Ms.*, *Pittsburgh's Child, South Florida Parenting, Connecticut Parent, North Jersey Parent & Child, Seattle's Child,* and *Kids Today* (Gannett Co., Inc.; for syndication information, call 800-368-3553).

Manufactured in the United States of America
10 9 8 7 6 5 4 3 2 1

Library of Congress Cataloging-in-Publication Data
Cella, Catherine.
Great videos for kids : a parent's guide to choosing the best / by Catherine Cella.
p. cm.
"A Citadel Press book."
Includes index.
ISBN 0-8065-1377-2
1. Video recordings for children—Catalogs. I. Title.
PN1992.945.C45 1992 92-31006
016.79143'75'083—dc20 CIP

To my guys, JERRY and MATT

ACKNOWLEDGMENTS

I would first of all like to thank the creators of the videos reviewed here. You have done marvelous work in a field that, sadly, attracts the mediocre. Thanks for your time and your talents . . . you've made my job a joy. Also I'd like to thank the myriad of public relations specialists who have supplied me with review videos and allowed me to donate them to children's hospitals, child-care centers, schools, and libraries.

For helping me track down titles in my quest to stay current and complete, I am indebted to Anisa Allen at Fusion Video. And for their generous research work, I thank the Putnam County (Tennessee) librarians, especially Kathleen Glinski, Reilly Reagan, Pat Wilhelm, and Wanda Maxwell. For helping me get children's input on videos, I am grateful to Mary McDevitt, Karolyn Cella, Gale Bennett, Kathy McEver, Pat Brown, Marsha Miller, and especially Paulette Witt, director of the child-care center which screens videos for me.

As my first young previewer, my son Matt deserves thanks, too. For holding the hand of a technophobe as she struggled with word processing, I owe thanks to Barry Stein, Ada Haynes, Dean Carothers, and my husband Jerry Neapolitan. And finally, I feel deep gratitude toward my mother and father and the teachers who instilled in me a love of art, including language. More than just informing this book, it's a love that enriches my life. Thank you.

CONTENTS

FOREWORD

BY SHELLEY DUVALL, PIONEER IN CHILDREN'S VIDEO

Rocky & Bullwinkle, Dumbo, Jiminy Cricket, The Tin Man, and more—all of my favorite characters are in these pages. There's so much great entertainment for kids, it has become increasingly difficult to sort through. And now, at last, there's a "roadmap" to the best that's available.

Television is an integral part of our culture, and if you've got kids, you will welcome this guide to viewing choices. Experience undoubtedly has the greatest impact in learning, but allegory has always been a means of teaching cultural values and customs from the early oral roots of storytelling through folklore and fables, and classic morality tales. This tradition has been carried through literature, theater, and the performing arts, to modern-day film, television, and now video. The good news is that home video gives us the flexibility to set our own individual agendas in terms of what and when we watch.

Catherine Cella is a fine reviewer who has applied her high, film critic standards to children's videos. With so much to choose from, her well-researched book is an invaluable guide to making informed choices. The titles are organized for easy reference—everything from your favorite stories and fairy tales to instructional videos, sing-a-longs, and even programs which deal with topical and sensitive subject matter. Television viewing can be a positive and affirming experience for kids of all ages (as well as their parents), and this book offers invaluable insights into a treasure trove of quality video titles.

Many engaging, engrossing and even challenging programs are listed in these pages. **Great Videos for Kids** is the definitive reference guide to responsible (and fun) video viewing. Make the most of what's out there!

Actress Shelley Duvall is executive producer of *Faerie Tale Theatre,* a multi-award winning series including the George Foster Peabody Award. Winner of seven Awards for Cable Excellence (ACE), she has produced *Shelley Duvall's Tall Tales and Legends,* also multi-award winning, and is executive producer of *Shelley Duvall's Bedtime Stories.* Ms. Duvall is in the Video Hall of Fame as an innovator in video programing. She also has her own production company, Think Entertainment, which is designed to develop, finance, and produce quality programing for children.

INTRODUCTION

Kids + TV is often considered a negative equation. Children watch too much television, and too much of what they watch is age-inappropriate, violent, or biased against women or minorities. And let's not forget commercials.

No wonder parents increasingly look to home video for their children's viewing. As ever-ready quality entertainment, kidvid can turn *Kids + TV* into a positive equation. But like any children's product, videotapes need careful screening with an eye to their impact on children and children's development.

This book will enhance your family's discretionary use of television. When watching its recommended videos with your children, you can encourage active, critical involvement. And you can build their media literacy by discussing how scenes fit together, why characters behave as they do, and how their actions have consequences.

Fortunately, there are many fine children's videos you'll *want* to watch with your children. *Great Videos for Kids* reviews such tapes, organized by areas: Animation, Book-Based, Educational, Family Topics, Folk and Fairy Tales, Holidays, Instruction, and Music. With the exception of animated features, movies aren't included because those reviews would fill another book.

To be selected, these Great Videos first had to meet professional standards of audio and video production. They also had to have little or no violence, racial stereotyping, or gender bias.

I quite naturally have some biases of my own. For one, I value beauty and am happily surprised at the quality of artwork in many of these videos. Having led children in song, I am similarly impressed with their musical excellence, not only in singalongs but as an integral part of many of the stories.

I further believe the best entertainment is educational in that it enriches the audience in some way. We all have favorite childhood stories that sparked our imaginations, excited our intellects, or touched our emotions—sometimes all three!

Finally, I am a children's advocate. I love kids and respect their opinions. So I share the videos I review with children at a child care center and in home settings. The videos reviewed in this book are those that entertained both me and my young helpers.

I trust you and your children will find enough treasures here to make this a valuable guide for years.

Catherine Cella

HOW TO USE THIS BOOK

The videos reviewed here are organized into categories for handy reference. If you want to see which books have been translated to video, you would check the Book-Based section. If your child loves music, there's a whole section devoted to sing-along, concert, and instrument-instruction videos. Or if you're looking for a particular folk or fairy tale, you would check that section.

Of course, many categories overlap considerably. Some videos could have been placed in any of several chapters. When checking for a specific title, you may want to consult the Title Index first.

Each listing begins with the title, followed by parenthetical information on the supplier, running time, and age range, and consists of a synopsis and review. You won't find price information because some videos have no suggested retail price, and those that do are often available at a discount. Suffice it to say, the average thirty-minute tape retails for around $15.00, a price point that is continually being lowered.

Some titles are grouped together for easy reference, such as those under **Dr. Seuss** in the Book-Based section or under **Peanuts** in Animation. Like books, videos go in and out of release. Some videos are included here because 1) they remain on library and video-store shelves, and 2) they may be re-released at any time. Be aware that the titles of some videos change as they acquire new emphasis, packaging, or ownership.

The video supplier, too, is variable. As companies fold and distribution arrangements end, home-video rights shift from one supplier to another. The situation often confuses even those in the business, coping with a video's producer, company, distributor, and labels within labels. The information here is accurate as of press time.

Finally, the age range given is my judgment—which doesn't always agree with the suggestion on the box—based on my knowledge of children and their responses to the videos. However, kids develop at different rates and in different directions. Please consider the age range a loose guideline to supplement your knowledge of your own children.

The video titles preceded by a star offer something special. The program may be particularly moving or beautiful or humorous or thrilling. Think of them as *A+* videos that merit your special consideration.

Great Videos for Kids

ANIMATION

Who doesn't love animation? Cartoons bring magic and laughter to all ages, and the best are evergreen, able to be enjoyed again and again.

Most of the videos reviewed here are done in traditional cel animation, where drawings are painted on Celluloid. Others are prime exmples of stop-action animation, where objects are moved with each camera shot to create the illusion of the objects moving on their own. And a few videos use computers for animation that has its own magic.

Fine animation is also included in the remaining seven sections, where it is noted in the reviews.

*__Adventures of Rocky and Bullwinkle, The__ (Walt Disney, 40–50 min., ages 6–14) These thoroughly-restored cartoons are, as billed on the video box, Classic Stuff. Like its descendants *The Simpsons* (the J in Homer J. Simpson is a nod to Bullwinkle J. Moose), *Rocky and Bullwinkle* has its own slant on things. Creator Jay Ward's offbeat sense of humor ranges from satire to spoof to so-dumb-it's-funny. In addition to an episodic Rocky and Bullwinkle adventure, each video includes minicartoons from the original program, such as *Fractured Fairy Tales* and *Mr. Peabody.* Classics among the classics are:

Volume 1—Mona Moose *The Treasure of Monte Zoom* is fast-paced and funny, funny, funny. Just when you think you have its humor pegged as off-the-wall, it goes through the fourth wall to poke fun at itself. Asides to the audience lampoon the clichés of both melodramas and cartoons. Side stories (*Fractured Fairy Tales, Mr. Peabody,* etc.) are so-so. But the adventures of our plucky heroes, as they race villains (Boris and Natasha) for treasure, are pure gold.

Volume 7—Whistler's Moose In one adventure, those no-goodniks Boris and Natasha sponsor a contest on "Why I Like Evil" in 25 words or less. However, Bullwinkle thinks it's "weevil," setting off an improbable chain of events. The second story has the island of Moosylvania being fought over by the United States and Canada, each pushing it back to the other side.

Adventures of Tin Tin: The Black Island, The (SVS, 40 min., ages 6–10) One of several videos animating the Herge comics of boy adventurer Tin Tin. With his pal Captain Haddock and dog Snowy, Tin Tin investigates mysteries surrounding long-lost treasure, castle ghosts, and shooting stars. In *The Black Island,* his adventures involve a ring of counterfeiters and a cave that fills with each tide. The animation isn't great, but Herge's uncluttered drawings and stories to match are engaging. Plus, Tin Tin thinks out loud with flashback images, so young viewers can share his problem-solving. Other good titles in this series include *The Secret of the Unicorn* and *Red Rackham's Treasure.*

An American Tail (MCA, 80 min., ages 3–8) Credited with sparking an animation revival, *Tail* relates Fievel Mousekewitz's search for his immigrant family. His story has humor, pathos, music, and top-notch animation. Producer Steven Spielberg gave animator Don Bluth a free hand, and it shows. Not only is the mouse tale credible, it's enhanced by such magical touches as the sparkle of water rings and cascading glass beads.

Bambi (Walt Disney, 69 min., ages 3–10) Featuring some of Disney's finest animation, *Bambi* also boasts such memorable characters as Thumper and Flower in addition to the fawn himself. His first experiences—walking, talking, rain—tickle his woodland friends as well as viewers. Bambi's story is moving, and the forest scenes breathtaking. However, some young children may be upset at Bambi's mother dying and frightened by the dramatic forest fire.

Ben and Me (Walt Disney, 26 min., ages 6–10) A 1953 Disney short claiming a mouse to be responsible for Ben Franklin's sayings, inventions, and news publishing. The cute story turns dramatic when Amos becomes the first high-flying reporter during a certain lightning storm. With excellent animation, this video makes good early education about our country's foundation.

***Betty Boop, The Best of —Volumes 1 + 2** (Republic, 85 min., ages 3–10) Max Fleischer's vamp debuted in the early, heady days of animation, and the joy of it all shows in every frame. A lightning bolt sews up the tear it made in the sky, an evil pirate reveals himself to be a rattlesnake, and penguins shape-shift as they dance down a chorus line. Beautifully restored and hand-colored, each video includes a dozen cartoons of the boop-boop-ee-doop heroine.

Bugs Bunny in King Arthur's Court (Warner, 25 min., ages 6–10) Chuck Jones animation of a story "stolen from Mark Twain." When Bugs pops up in medieval times, he must convince folks that he's not a fire-breathing dragon but a carrot-eating rabbit. With his 'toon friends, Bugs tweaks the original story with a few jokes of his own. Well-drawn, colored, and animated, this *Court* is supreme fun.

Bugs Bunny's Hare-Raising Tales (Warner, 45 min., ages 4–10) Six takeoffs cast Bugs in such classics as *Aladdin* and *Robinson Crusoe.* Funniest spoofs include *Knight-Mare Hare,* with its twists of medieval language and lore. And in *Rabbit Hood,* the Sheriff has met his match in the carrot-stealing Bugs. Despite its standard bops on the head, the cartoon has enough wordplay to earn its laughs.

***Cat Came Back and Three Other Titles, The** (Smarty Pants, 30 min., ages 4–12) National Film Board of Canada animation, including Cordell Barker's treatment of the title 100-year-old folk song. The Oscar-nominated cartoon follows a man's desperate—and incredibly unsuccessful—attempts to rid himself of the infamous cat. Wittily drawn and animated, this *Cat* attracts all ages. Also included are *What on Earth, The Egg,* and *Blackberry*

Subway Jam, a watercolored tale of a boy investigating why his room suddenly became a subway stop.

Chip 'n Dale Rescue Rangers (Walt Disney, 44 min., ages 4–10) Each video in this line includes two episodes of the Indiana Jones-type TV program. The little fellas take on such adventures as finding a missing cat and battling pi-rats for sunken treasure. Well-animated and scored, *Rescue Rangers* is inventive, funny, and full of the charm of small ones coping with an outsized world.

Chipmunk Adventure, The (Warner, 76 min., ages 3–10) Simon, Theodore, and Alvin's first feature film pits them against the Chipettes in a global balloon race. Dramatic without being scary, the story has its share of fun as well. And there's the bonus of international flair as they travel to Greece, Egypt, and Mexico. Colorful and active without violence, this *Adventure* appeals to a wide age group.

Chuck Amuck: The Movie (Warner, 51 min., ages 8–14) A documentary on animator par excellence Chuck Jones. Interview segments, where he talks about his work and childhood influences, are cut with illustrative cartoon clips. The seven-minute animated short, *What's Opera, Doc?*, finishes the tape. An interesting look at animation and its creation of living characters.

Computer Dreams (MPI, 60 min., ages 8–14) PBS special showcasing the burgeoning art of computer animation. Claiming it's like taking a camera into your dreams, the video delivers with fantastic landscapes, statuary coming to life, and surreally-smooth geometrics. A nice balance is struck between showing and explaining. Best segments reveal the process of creating a digital actor and the emergence of a building from its blueprint. Hi-tech magic.

Denver the Last Dinosaur (Fries, 45 min., ages 5–10) Recommended by the National Education Association, *Denver* combines the colorful action of Saturday cartoons with good stories that educate. In this premiere episode, some teens discover an egg in the La Brea Tar Pits. When Denver is hatched, his shell magically transports the kids back to prehistoric times. Returning to our time, the shades-wearing and skateboarding Denver fits right in, even if he is "an odd-looking kid." Other tapes offer good, nonviolent adventures as well.

Dingles and Three Other Titles, The (Smarty Pants, 30 min., ages 3–8) Title tale, in award-winning Les Drew animation, tells of Doris Dingle and her three idiosyncratic cats. One is a Siamese snob, another gets into curious mischief, and the third insists on digging a hole to China. Humorously drawn and animated, they do everything together—from enjoying tea time to saving each other from a windstorm to sharing an indoor picnic. Of the three remaining tales, best is *The Boy and the Snow Goose.* An award-winning tale of affection and letting go, it's told with beautifully simple music and art.

*__Donald's Bee Pictures__ (Walt Disney, 49 min., ages 3–10) A Limited Gold Edition video that may be available only to rent. The irascible duck is pitted against his nimblest nemesis in seven cartoons. Cat and mouse were never so funny as duck and bee in their attempts to outwit each other. The animation is excellent and as clever as our two adversaries. A honey of a video worth seeking out.

Dot series This Australian line stars a clever animated girl named Dot against a live-action background. Her love of animals sends her into adventures that captivate creature-loving kids. They're educational, too, about Aussie animals, their habitats, and ecology. The best Dot adventures are . . .

Dot and the Bunny (CBS-Fox, 79 min., ages 4–8) Our heroine sets off to find a mother kangaroo's joey. Her trek through Australia turns up many of its unusual animals in funny and educational scenes. One scene, however, may frighten younger viewers, as a crocodile sings about nibbling Dot's fingers and bones.

* **Dot and Keeto** (FHE, 70 min., ages 3–8) Like an Australian Alice, Dot shrinks to insect size and befriends Keeto the mosquito. In her quest for the root to make her big again, she

meets insects good, bad, and ugly. All have their place, she learns, through animated characters and fascinating nature footage. A good blend of live action and animation, story and song, make this video as entertaining as it is educational.

Dot and the Whale (FHE, 75 min., ages 3–8) It's underwater adventure for our sweet heroine, as she tries to help a beached whale. In her travels she encounters sharks, penguins, poisonous coral, and the ice cave of Moby Dick himself. She learns that it's up to humans to protect the great creatures from whalers. Actual footage of leaping whales adds force to the plea.

Ducktales: Daredevil Ducks (Walt Disney, 44 min., ages 6–11) One of a series collecting two episodes each of the top-rated Disney show starring "Unca" Scrooge and nephews. As in the comic books, their adventures are inventive, funny, and nonviolent. And they're better animated than most TV cartoons. *The Money Vanishes* features a transporting gizmo that falls into the wrong hands, the Beagle Boys. And *Home Sweet Homer* finds Scrooge and the boys helping the Greek king oppose Circe's evil magic.

Ducktales: The Movie—Treasure of the Lost Lamp (Walt Disney, 74 min., ages 4–9) Like the TV series, this movie offers fun, mile-a-minute adventure à la Indiana Jones. Scrooge and the kids explore a pyramid filled with as many traps as treasures. Thanks to Merlock, an evil sorcerer voiced by Christopher Lloyd, all they emerge with is a lamp. Luckily, and sometimes unluckily, for them the lamp holds a young genie. His mix-ups and the ducks' cleverness in opposing Merlock are the most entertaining parts of the film.

Dumbo (Walt Disney, 63 min., ages 3–7) The little elephant with the big ears moves from trials to triumph in this Disney animation aimed at the very young. At first teased and prodded into becoming a clown, Dumbo finds the courage to fly with a little help from his crow friends. What the film lacks in complex animation it makes up in cute characters and simple circus fun.

Dumbo's separation from his mother, however, may disturb some younger viewers.

*__Every Child and Four Other Titles__ (Smarty Pants, 30 min., ages 6–10) These National Film Board of Canada animated shorts open with an Academy Award-winner. Dedicated to UNICEF, it conveys simply and humorously the right of every child to a home. *The Magic Flute,* with its lyrical and childlike art, creates more than music. Then the McGarrigle Sisters sing the rousing *Log Drivers' Waltz*, recalling the old-time birlers who drove logs downriver to the mill. Rounding out the tape are *The Town Mouse and the Country Mouse* and *Catour,* starring a jazzy feline. An animation tour de force.

*__Fantasia__ (Walt Disney, 2 hrs., ages 3–12) In 1941, Walt Disney's visionary masterpiece failed to find its audience. Fifty years later, *Fantasia* sold over fourteen million copies to become the all-time best-selling video. The idea was born when it became apparent that *The Sorcerer's Apprentice* short wasn't going to pay for itself. So Disney created a feature by adding such pieces as the hippo slapstick of *Dance of the Hours, The Rite of Spring's* dinosaur dramatics, the abstract grace of Bach's *Toccata and Fugue,* and *The Nutcracker Suite's* nature fantasy. Disney's dream of *Fantasia* as a living project with regular updates comes true in 1996 when *Fantasia Continued* is projected for release.

Felix the Cat: An Hour of Fun (VidAmerica, 60 min., ages 5–9) Better than *Felix the Cat: The Movie* is this collection of adventures from the 1960s TV show. The humor is corny-camp and the animation only fair. But there's a charm to Felix, with his belly laughs and preposterous bag of tricks, that's hard to resist. Also available is *Another Hour of Fun.*

Fireman Sam: The Hero Next Door (FHE, 30 min., ages 3–6) Like *Thomas the Tank Engine and Friends*, *Fireman Sam* is British puppet-mation that emphasizes helping and friendship. With an array of voices, John Alderton narrates the stories of our hero of Pontypandy town. Retrieving a kite from the rooftop is as exciting as Sam's adventures get. They're as simple and engaging as those a child might invent while playing with these toylike characters.

Flintstones: A Page Right Out of History, The (Hanna-Barbera, 30 min., ages 3–10) 1991 made-for-video celebration of the first prime-time cartoon's thirtieth anniversary. Like most tribute/history programs, it combines clips with interviews. Hanna and Barbera obviously still get a kick out of the Flintstones and their particular humor. Included is such trivia as how "Yabba-Dabba-Do" originated, but curiously absent is any mention of the inspiring Honeymooners. Tape ends with the episode of Pebbles's arrival, "a page right out of history" as the first cartoon baby birth.

Flintstones: The First Episodes, The (Hanna-Barbera, 96 min., ages 3–10) Like the Honeymooners characters they're based on, Fred and Barney draw humor from slapstick, fat jokes, bungled schemes, and predicaments. The episodes here involve their lodge talent show, a shared swimming pool, Fred's losing Barney's job, and a fabulous Flintstone flying machine. It's a fun, upbeat collection of the best of the Modern Stone Age Family.

Fun on the Job (Walt Disney, 30 min., ages 3–8) Three of the four animations on this tape are indeed Cartoon Classics. In 1941's *Baggage Buster,* Goofy takes on a magician's trunk and all its tricks. *The Clock Cleaners* (1937) are Mickey, Goofy, and Donald, who gets stuck in the mainspring of a tower clock. And *Mickey's Fire Brigade,* from 1935, has the gang in early form as they battle a blaze. Beautifully drawn and colored, *Brigade* is also funny, as the flames take on a life of their own.

Garfield on the Town (CBS-Fox, 24 min., ages 3–10) The fat cat's Emmy Award-winning special opens with his usual antics with Odie. They cause him to fall out of the car and into a strange neighborhood. Adrift in the big city, Garfield rumbles with a gang and finds his roots in an Italian restaurant. It's a good story packed with jokes about Garfield the housecat—part house, part cat. See also *Garfield Christmas Special.*

Here's Donald! (Walt Disney, 25 min., ages 3–8) A Cartoon Classic of three cartoons starring Donald Duck. In *Wide Open Spaces,* he tangles with the not-so-great outdoors. *Donald's Ostrich* swallows a radio and enacts whatever it announces—a fight, a race, etc. And in *Crazy With the Heat,* Donald and Goofy hallucinate the Sahara having the wildest mirages. All feature creative animation, funny sight gags, and hijinks of the irrepressible Duck.

Here's Mickey! (Walt Disney, 27 min., ages 3–8) Cartoon Classic of three cartoons starring Mickey Mouse and friends. For *Mickey's Birthday Party,* Donald Duck "hides" himself as a lamp and steals the show with a Mexican dance. In *Orphan's Benefit,* Mickey and the gang put on a vaudeville show, hook and all. And *Mickey's Garden* shows the Mouse in early form, trying to rid his garden of pests. As fresh today as ever, each *is* a classic Disney 'toon.

Journey Through Fairyland, A (Celebrity, 95 min., ages 6–10) Owing a clear debt to *Fantasia,* this animation gives visual expression to over twenty classical compositions performed by the Tokyo Philharmonic. When young Michael must choose between his two loves—music and flowers—he shrinks to sprite-size for an adventure in his garden. This makes for some beautiful and dramatic scenes as he dances amid flowers and faces menacing insects. Despite some off notes in the animation and plot line, this *Journey* is well worth taking.

Lady and the Tramp (Walt Disney, 77 min., ages 3–9) Disney classic that's as strong technically as it is emotionally. Four years and $4 million in the making, the film has such memorable scenes as our lovers eating spaghetti on a "Bella Notte" and the feline mischief of the "Siamese Cat Song." Scored by Peggy Lee with Sonny Burke and wonderfully animated, Lady and the Tramp's story has a fine supporting cast of canine character "actors."

Land Before Time, The (MCA, 69 min., ages 2–9) Don Bluth animation of dinosaur babies on their own. Separated from his mom during an earthquake and Tyrannosaurus rex attack, Littlefoot makes his way to the Great Valley. He and his new friends encounter all the beauty and drama of prehistoric times. Well-written for the very young, the story may also upset some as it springs from the death of Littlefoot's mother. Still, the baby dinosaurs are irresistibly cute and their quest nicely animated.

***Life With Mickey!** (Walt Disney, 47 min., ages 3–10) A Limited Gold Edition video, this may be available only as a rental. Its seven cartoons showcase not only the most famous mouse in the world, but also the joys and progress of animation. Opening with a brief bio of Mickey, the video's best cartoons have him sharing the stage with Donald Duck. In *Alpine Climbers,* Donald's tussles with a baby goat are priceless. Worth seeking out.

Little People videos (R&G/Fisher-Price, 30 min., ages 2–6) This animation of characters from the Fisher-Price toy line casts them in sweet little stories for the very young. Simply told and artfully drawn, the videos are heartwarming without being sappy. Some of the best in the line are:

Little People: Christmas Fun —see Holidays: Christmas

Little People: Favorite Songs —see Music

Little People: Jokes, Riddles, and Rhymes The family must amuse itself when stranded by a rainstorm in the local theater. They tell riddles and jokes about everything from food ("What kind of shoes are made from bananas? Slippers!") to ghosts who eat Scream of Wheat. Acting out classic nursery rhymes completes this entertaining video for young ones.

Little People: Three Favorite Stories As the family reminisces over a photo album, stories emerge about cousin Marvin's robot, Penny and Timmy's baby-sitting, and Lucky's dog contest. Praising imagination and kindness, each ends with a moral—such as helping your friends is more important than winning.

Lollipop Dragon: Magic Lollipop Adventure, The (Celebrity, 25 min., ages 2–6) Baron Bad Blood has stolen the magic lollipop and three baby dragons from TumTum. To the rescue comes the Lollipop Dragon, who with his friends must face the Forest of Mirrors and Blood Castle. Kids love that the babies have magical powers. You'll appreciate that the conflict is nonviolent and cleverly resolved in restitution. A colorful, imaginative story that shows friends cooperating in their quest.

Maple Town: The House Made of Love (FHE, 45 min., ages 5–9) If you can forgive this animated series for being toy-related, you'll find its messages promote kindness and fair treatment. In the title story, a gingerbread house helps Kathy realize she's become the kind of grown-up she doesn't like. In the second tale, *A Most Unlikely Heroine*, the Maple Town Players hit a few snags in their latest production. Both are good stories and well drawn, treating human foibles in a context young children can understand.

Mister Magoo in Sherwood Forest (Paramount, 80 min., ages 5–12) Opening with Magoo's demonstration of the flaw in Robin Hood's plan—that is, who The Rich are changes with each transfer of gold—the video moves into a surprisingly straightforward retelling. Magoo plays Friar Tuck and narrates the story of how the Merry Band came together and became outlaws. A well-written and animated introduction to the legend.

Nuts About Chip 'n Dale (Walt Disney, 22 min., ages 3–10) Three early cartoons featuring the antagonistic antics of today's Rescue Rangers. *Two Chips and a Miss* has Chip 'n Dale vying for the attentions of a nightclub singer. *Food For Feudin'* pits them against Pluto and his bone burying. Best of the lot is *Trailer Horn*, with its classic mischief focused on Donald Duck.

*__101 Dalmations__ (Walt Disney, 80 min., ages 3–12) One of Disney's finest—an expert blend of comedy and drama in a great rescue story. When Pongo and Perdy's litter become part of 99 pups in Cruella De Vil's fur-coat plans, the hound alliance mobilizes. Courage, cunning, and cooperation all figure in the rescue, as do some of the best canine characters on the screen. The animation is smooth and story well paced for virtually all ages.

Peanuts videos If Charles Schulz did nothing else, what he could express with one simple mouth line would ensure his genius status. But then, he also created some of pop culture's most beloved characters. And his perception of the human condition, as viewed through the *Peanuts* prism, helps us laugh at ourselves and *like* it. The best of his video adaptations . . .

Peanuts: Be My Valentine, Charlie Brown —see Holidays: Valentine's Day

Peanuts: Charlie Brown's All Stars (Lee Mendelson, 30 min., ages 3–10) "We're doomed!" say the boys, girls, and dog of Charlie Brown's baseball team. Their luck turns for the better when Linus finds a uniform-buying sponsor. But when the league disallows girls and dogs, the All Stars take a new direction. Well-drawn and written, this baseball classic is at turns funny, poignant, and endearing.

Peanuts: It's a Mystery, Charlie Brown (Lee Mendelson, 26 min., ages 3–7) It's The Case of the Missing Nest, as Sherlock Snoopy searches for Woodstock's home. Judge Lucy presides in a courtroom full of legal mumbo jumbo. And Sally Brown finds herself in quite a spot. Lean on dialogue, this video's animation virtually tells the story, making it perfect fare for young detectives.

Peanuts: It's The Great Pumpkin, Charlie Brown — see Holidays: Halloween

Peanuts: It's Your First Kiss, Charlie Brown —see Holidays: Valentine's Day.

*__Peep and the Big Wide World__ (Smarty Pants, 30 min., ages 3–10) This modern fable, narrated by Peter Ustinov, tells of innocence gradually informed but never daunted. Peep is a newborn chick "so young she doesn't know what feet are for." She hooks up with a cranky duck named Quack and the world-wise robin Chirp. Together they make their way in the Big Wide World, conquering adversity and dancing when necessary. The animated art is wonderfully simple and storytelling flecked with humor and insight.

***Pegasus** (Lightyear, 25 min., ages 5–10) As the voice of the youngest muse, Urania, Mia Farrow tells the winged horse's story with youthful excitement. When Athena learns Urania cannot sing, she sends her Pegasus as a gift. Their play and growing affection make fine counterpoint to the dramatic struggles which follow. Beautifully written, scored (by Ernest Troost), drawn, and animated, *Pegasus* soars. It's filled with monsters and magic, making it a great introduction to Greek classicism.

***Peter and the Wolf** (Walt Disney, 30 min., ages 3–6) This "Fairy Tale With Music" makes a wonderful intro to classical music. Each character in the Prokofiev score has a representative instrument-motif. A string quartet plays Peter, the bird is a flute, duck an oboe, and so on. Sterling Holloway narrates Peter's adventure along with the music in this riveting production. Also included on the video are *Symphony Hour,* in which Mickey Mouse conducts a cacophonous orchestra, and *Music Land,* a Silly Symphony of the discord and eventual harmony between the Land of Symphony and the Isle of Jazz.

Puff the Magic Dragon (Vestron/FHE, 40 min., ages 4–8) Burgess Meredith narrates this original and best of the Puff cartoons. The dragon's magic puffs reveal the reality behind the front. A giant pirate, for example, is shown to be a would-be chef. As in the song, Puff takes little Jackie Draper to the land of Honilee, where he overcomes his fear of talking. There they sing songs and discover the magic of friends helping each other. Nicely drawn and animated, *Puff* sings a song of imaginaton.

Railway Dragon, The (FHE, 30 min., ages 4–10) Suspecting there's magic within an old railway tunnel, Emily ventures in and discovers it's so. An old dragon lives there with his "treasure" of items that have fallen off trains. After he tells her about dragons being near extinction due to humans' fear, she talks him into attending a dragon celebration. There they see dragons emerge from their camouflage in tree gnarls and roots. And Emily shares in their revelry of limbo, card games, and self-heated popcorn.

This original story is treated to fine animation and conveys subtle messages of tolerance and the power of friendship.

*__Red Shoes, The__ (FHE, 30 min., ages 6–10) An update of the Hans Christian Andersen concept into an urban tale of two friends. One's sudden wealth changes her for the worse until a whirl in the red shoes sets her straight. Ossie Davis narrates, with sunny Caribbean music adding its own rhythms to the story. As usual, Michael Sporn's animation is a work of art, this time in bright, shifting pastels.

Reluctant Dragon, The (Walt Disney, 28 min., ages 3–10) Kenneth Grahame's story is treated to fine and funny animation from Disney. Reluctant to fight Sir Giles, a poet dragon enlists the help of a boy. Together the three hatch a plan to fool the public, where the smoke of "battle" hides what's really a tea party. The characters are well-drawn quirks, and the message of nonviolence comes through clearly but with a light touch. Also included is *Morris the Midget Moose,* a funny tale of cooperation where Morris teams up with a complementary friend to take on Thunderclap for leadership.

Rescuers Down Under, The (Walt Disney, 77 min., ages 3–10) With over a million drawings and three years in the making, Disney's twenty-ninth animated feature shows nobody does it better. Mice Bernard (Bob Newhart) and Miss Bianca (Eva Gabor) tangle with nasties in the Australian outback. And the action, including some fantastic flying swoops, is nonstop and nonviolent. The only drag is, appropriately, the decidely unfunny albatross (John Candy). Fortunately, the Rescuers save the film with their cute and daring antics.

*__Rupert__ (CBS-Fox, 57 min., ages 4–9) While not truly animated, this video's artwork and storyline are strong enough to bring the British bear to life. Over there, Rupert has been a comic-strip favorite for over seventy years. This video shows why, with a dozen tales of mystery and magic. Rupert and his chums investigate the likes of ice flowers and sea sprites in a world of childhood fantasy. Also available, though not as entertaining, is *Rupert and the Runaway Dragon.*

***Rupert and the Frog Song** (FHE, 22 min., ages 4–9) Produced by Paul McCartney, this animation of the British bear is Disney-esque in its flowing quality and magical touches. After passing shimmery waterfalls and autumn leaves that become butterflies, Rupert stumbles upon a froggie festival. "We All Stand Together," they sing and dance and play in scenes reminiscent of *Fantasia.* Also included are two Linda McCartney animations—the unusual, abstract *Oriental Nightfish* and *Seaside Woman,* the imaginative interpretation of her reggae song that won top honors at Cannes.

Salute to Chuck Jones, A (Warner, 56 min., ages 3–10) A tour de force from Warner's most inventive animator. The eight cartoons included range from a space spoof starring Daffy Duck and Porky Pig to a song-and-dance frog absurdity to *High Note,* where sheet music is too playful to be playable. A fitting climax is the classic *What's Opera, Doc?*, puncturing Wagnerian grandeur with Warnerian folly. Elmer and Bugs star in this and other cartoons, in addition to stories of Pepe Le Pew, Road Runner, and Wile E. Coyote.

***Scamper the Penguin** (Celebrity, 75 min., ages 4–10) Just as cute as it sounds. Opening with bits of education about penguins, the video moves on to the adventures of young Scamper and his girlfriend Snowflake. Their explorations lead to trouble, but they find their way home with courage and cooperation. The first joint project of leading animators from Russia and Japan, *Scamper* is a minitrip to Antarctica in all its beauty and drama.

Scooby and Scrappy-Doo, Volume 2 (Hanna-Barbera, 60 min., ages 6–10) Three mysteries send Scooby and the gang off to investigate a living comic-book hero, a flying saucer, and the theft of the crown jewels. Each is a good puzzle, with clues throughout which are then reviewed near the end. With only mild violence and plenty of humor in Scooby's patented double take, this video is worth seeking out, if not in current release.

Sea Dream and Two Other Titles (Smarty Pants, 30 min., ages 6–10) Video collection of award-winning National Film Board of Canada animation. Title short follows a girl's dream where she plays baseball and has tea with an octopus. *Nebule* stars a magic cord that can become stairs, a tightrope, a monster, anything. And *The Sound Collector* is a boy who creates new meanings for familiar sounds. Is that a rocking chair or a robber with squeaky shoes? All three stories celebrate the imagination of children.

***Silly Symphonies** (Walt Disney, 60 min., ages 3–8) Vintage Disney of seven cartoon musicals. *The China Shop* comes alive at night when the satyr kidnaps the woman of a pair of loving figurines. The ensuing battle is filled with clever animation. In *The Cookie Carnival*, a ginger boy decorates his girlfriend to become queen of the parade. With a vaudevillian song-and-dance finale, it's pure confectionary delight. And *Woodland Cafe* peeks in on a miniature nightclub where fireflies light the tables and real jitterbugs dance the night away. If you can still find this older release, grab it.

Sports Cartoons (FHE, 50 min., ages 5–10) A fun little video collecting animated vignettes of sporting animals. One-to-two-minute segments star the likes of basketball-dribbling cats and parachuting hippos. Artfully drawn and shot with sophisticated humor, these sports spoofs hold appeal for adults as well as kids.

Starring Chip 'n Dale (Walt Disney, 22 min., ages 3–10) A Cartoon Classic of three shorts starring the little bedevilers. In *Working For Peanuts,* they prey on a circus elephant's feed, to the dismay of trainer Donald Duck. In *Donald Applecore,* they make applesauce out of the duck and his orchard. And *Dragon Around* finds them suited up in the most inventive armor as they battle Donald's dragon. Solid Chip 'n Dale fun.

Starring Mickey and Minnie (Walt Disney, 27 min., ages 3–9) Three cartoons from the Disney vaults. *The Little Whirlwind* (1941) has Mickey tangling with a mischievous minitornado in a classic Mouse 'toon. *Hawaiian Holiday* (1937) finds Mickey playing ukulele, Pluto chasing starfish, Goofy surfing, and

Donald hula dancing. And *The Brave Little Tailor* (1938) pits Mickey against a giant to win the hand of Princess Minnie. Using the tools of his trade, the Little Tailor manages to subdue the giant in this fine animation of the medieval tale.

Teddy Bears' Picnic, The (FHE, 30 min., ages 4–8) For one magical day each year, teddy bears come alive to party in the forest. It's young Wally's first time, so he forgets the rule and invites a sad little girl. Lost in the woods and without her teddy, Amanda perks up at the chance to attend this special picnic. A cute and original story based on the title song.

***Tender Tale of Cinderella Penguin and Four Other Titles, The** (Smarty Pants, 30 min., ages 4–10) Hilarious and marvelous adaptation, casting penguins in the classic fairy tale. This Academy Award nominee is set to medieval music without narration, so visuals tell the story. Picture penguins struggling into corsets, ballroom dancing, and losing a glass flipper, and you have some idea of *Cinderella Penguin's* delights. Of the four additional animations, standouts are the mesmerizing *Metamorphoses* and *The Sky Is Blue,* about a boy's flight of imagination.

BOOK-BASED

Reading a book with your child is one of parenthood's perks. Watching him or her learn, delighting in the artwork, sharing a story and, best of all, time together make it a special experience. This doesn't have to be unique to books, however, as children's video has some classics of its own. Not coincidentally, many are based on—and lead viewers back to—fine children's literature.

Abel's Island (Random House, 30 min., ages 4–10) Like *Doctor De Soto*, this is a Michael Sporn animation of a William Steig book. And again a mouse uses his wits to survive, this time on a deserted island. The animation is a delight, in things as subtle as the swish of a scarf or as dramatic as a summer rainstorm. And the story, philosophical and yet simple enough for young children, holds appeal for all ages.

Alice in Wonderland (Disney, 75 min., ages 5–12) Lewis Carroll nonsense meets Disney magic in this feature-length adaptation. Alice's search for the White Rabbit leads her to such crazies as the Mad Hatter, the Cheshire Cat, Tweedles Dee and Dum, and over them all, the Queen of Hearts. While at times Alice's frustration is all too real, Disney's gift for characterization and Carroll's wit marry well. The Mad Hatter's tea party is a gem.

Anne of Green Gables (Walt Disney, 199 min., ages 7–14) L. M. Montgomery's turn-of-the-century novel translates well in this adaptation starring Megan Follows, Colleen Dewhurst, and Richard Farnsworth. As the ultimate romantic adolescent, Anne finds poetry in nearly every event. Her wistfulness never overtakes her intelligence, however, in facing life's demands. And her eloquence in striving to be accepted has endeared the character to young readers for generations. Just as Follows brings Anne to life, this production recreates her world in such detail you hate to see it end—just as with a good book.

***Are You My Mother? And 2 More P. D. Eastman Classics** (Random House, 25 min., ages 2–5) With fine narration and complementary music, three P. D. Eastman stories are cleverly, if slightly, animated. The first is full of funny little surprises, as a baby bird determines to find his mother. *Go, Dog, Go!* shows canines of different colors, shapes, and sizes. And *The Best Nest* has a bird couple searching for a new home. All are excellent stories for the very young, with alliterative wordplay and art that is simple yet interesting.

Babar videos Few picture books hold the intricate charm of the Babar series, begun by Jean de Brunhoff and carried on by son Laurent. Generations of children and adults have pored over the books, sharing their wonderful lines, both drawn and written. Their translation to video has been hit-and-miss, depending on adherence to the originals' spirit. Here are the hits:

Babar and Father Christmas —see Holidays: Christmas

* **Babar Comes to America** (Vestron/FHE, 30 min., ages 3–6) This adaptation of the *The Travels of Babar* and *Babar Comes to America* benefits from Laurent de Brunhoff's script, Peter Ustinov's narration, and animation by *Peanuts'* producers Mendelson and Melendez. The King of the Elephant's entourage includes Queen Celeste, the wise old Cornelius, and rambunctious cousin Arthur. In their travels cross-country, elephants and humans mingle quite naturally. And in Hollywood, Arthur sets out to become a star. It's a winsome tale that successfully draws on the books' main attractions.

Babar Returns (FHE, 49 min., ages 3–6) Also includes *The City of Elephants*, both episodes from the HBO Babar series. While the artwork and stories are simpler than the original books, they do mirror Babar's positive values. In the first story, the elephants triumph over the hunter through cleverness rather than violence, and in the second they learn the wisdom of doing something well rather than quickly.

* **Babar the Little Elephant** (Vestron/FHE, 30 min., ages 3–6) Babar's origins are traced with care in this animated adaptation written by Laurent de Brunhoff and narrated by Peter Ustinov. His mother felled by a hunter's bullet, Babar runs off to the city where he lives with the Old Lady. Two years later, he returns to the forest where he's crowned King and builds Celesteville. When the town is threatened by attacking rhinos, Babar comes up with a nonviolent counterploy. The story rolls along with the simple grace of the books and finishes with an uplifting parade.

Babar: The Movie (FHE, 79 min., ages 3–6) The little elephant's first feature film tells the story behind the elephants' annual Victory Parade. When Babar was first King, the rhinos waged war on the elephants until Babar figured a way to outwit them. Based on the characters and not the books, this video leans to Saturday-cartoon style, with its emphasis on action and simplistic villain formula. Still, there are funny moments in Zephir the monkey's antics, and its values of cooperation and nonviolence are true to the original de Brunhoff text.

***Baby's Bedtime** (Lightyear Entertainment, 27 min., ages 2–5) Judy Collins is in fine and pretty voice, singing over a dozen lullabies in this Stories to Remember video. Kay Chorao's book of poetry receives wonderful treatment from Daniel Ivanick's animation and an Ernest Troost score. In the opening "Dance, Little Baby," a mother and infant waltz through a room of stuffed animals. "The Japanese Tree Shadows" stars a big yellow moon on a midnight blue sky. And "Hush, Little Baby" shows a father and child rocking to scenes from the song. Other delights include poems from Robert Louis Stevenson, Alfred Lord Tennyson, and Gaelic tradition. A warm and beautiful bedtime video.

***Baby's Morningtime** (Lightyear Entertainment, 25 min., ages 2–5) Beauty of a wake-up video based on Kay Chorao's *The Baby's Good Morning Book.* Poems by such authors as Robert Browning and Emily Dickinson are set to wonderful music by Ernest Troost and sung by Judy Collins. Animated in flowing, natural imagery, *Morningtime* has a rhythm designed to help little ones break into the day. Try it after naptime, too.

Baby's Storytime (Lightyear Entertainment, 28 min., ages 3–6) Stories to Remember video adapting Kay Chorao's *The Baby's Story Book.* Narrated by Arlo Guthrie and attractively animated by Michael Sporn, these traditional tales have been tailored for today's children. Of the dozen stories, best are *The Gingerbread Boy, The Princess and the Pea, The Wind and the Sun,* and the alphabetical *History of an Apple Pie.* Apropos to the material, this video has a nice, old-fashioned quality to it, with bits of leavening humor.

Baby-Sitters Club videos (Scholastic/GoodTimes, 30 min. each, ages 6–12) Like the wildly best-selling Scholastic book series, these video adaptations are simply wonderful. Sharing their lives as well as a business, seven girls resolve problems in an honest, upbeat way. And the storytelling is elegant, typically developing several plotlines at once. As the theme song goes, these girls can count on each other. And you can count on excellent entertainment from the Baby-Sitters Club.

Baby-Sitter's Club Special Christmas—see Holidays: Christmas

Claudia and the Missing Jewels The first Baby-Sitters Club mystery on video, *Jewels* opens with Claudia launching a jewelry-making business. When a pair of earrings is stolen, the girls have their suspicions, but are they right? While this episode is a bit oversweet, it's still good enterainment.

Dawn and the Dream Boy The girls may be rooting for Dawn to go to the Sweethearts Dance with Jamie, but he seems more interested in her stepsister Mary Anne. Friction between the two girls mounts until an emergency meeting of the Club must be called. Briskly paced and with party-planning scenes, the video nicely balances teen fun and troubles.

* **Dawn and the Haunted House** Is Claudia bewitched? That's what Dawn thinks when Claudia spends too much time with the neighborhood "witch." When learning the truth punctures her theory, Dawn also learns to appreciate different people

and the importance of not jumping to conclusions. Well-edited to build suspense toward a satisfying conclusion, the video's a classic good story well told.

* **Kristy and the Great Campaign** In her overzealous way, Kristy does a complete makeover for Courtney in her bid for third-grade president. A good vehicle for issues of self-knowledge and self-concept, *Campaign* counters gender stereotypes as well. Not only do both girls and boys scream at a snake on the loose, Courtney's the one who calmly picks it up to bring to science class.

Mary Anne and the Brunettes The situation: One of the Brunettes is flirting with Mary Anne's boyfriend, who has no idea why she is upset. The solution: At first it's the Club against the Brunettes, until finally the couple themselves work things out. With dialogue that rings true, this video explores a typical problem in a rational and entertaining way.

* **Stacey's Big Break** Stacey seems headed for modeling stardom. While this excites the girls about her future, Stacey's more concerned with what she's missing in the present. This sets up dialogue on important issues such as priorities and anorexia. And yet the program remains lively and entertaining. Beautifully lit and photographed, this is a memorable meeting of the Club.

Berenstain Bears videos (Random House, 20–30 min., ages 3–8) Stan and Jan Berenstain have been cleverly educating children since 1962. Their best-selling Bears resolve family problems realistically and with good humor. Add the most ingenious rhymes and you have a winning combination. The best of their animated adventures . . .

Berenstain Bears and Cupid's Surprise —see Holidays: Valentine's Day

Berenstain Bears and the Trouble With Friends —see Family: The Environment

* **Berenstain Bears and Too Much Birthday + To the Rescue** Sister is six years old and despite Mama's warning, it looks like Papa is overplanning her party. On top of games and cake, they have ponies, a merry-go-round, and helium balloons. In the

end, Sister is not the only one who's had too much. Second tale is patented Bear fun, as Papa helps the Bear Scouts earn their merit badges, but not in the way he intended.

Berenstain Bears' Christmas Tree —see Holidays: Christmas

Berenstain Bears' Easter Surprise —see Holidays: Easter

Berenstain Bears: Get in a Fight + The Bigpaw Problem Two stories about working out differences. In the first, Brother and Sister engage in typical sibling rivalry until Mama brings them together again. Her suggestion to stop and think about their disagreement before it escalates to a major rift is a good parenting lesson, too. She then reinforces it with a metaphor of a fight being a storm that will blow over if calm prevails. Second story has Bear Country resolving to accept the good and bad points of the oversized Big Paw. Both are nicely animated and paced for young children.

Berenstain Bears: In the Dark + Ring the Bell When Sister is frightened by Brother's scary book, she wants the bedroom light on at night. But Brother can't sleep with it on, leading to a good talk about imagination and a solution from Papa. Second story has the Bear family competing at the fair, including Papa's violet-and-onion flavored honey. Whether he'll be able to ring the bell in the Strongest Bear Contest is the focus of the story.

* **Berenstain Bears: The Missing Dinosaur Bone/Bears in the Night/The Bear Detectives** These slightly animated tales provide great intros to the Bear family *and* to detective fiction. In the first the Bear detectives and their hound Snuff investigate a museum's missing dino bone. They succeed by careful search, observation, and deduction. Similarly in the third story, Papa goes awry as usual, but the cubs solve the "crime" by patiently following tracks and other clues. Clad in deerstalker caps, the Bear Detectives offer good, simple mysteries for the youngest of PI's.

* **Berenstain Bears: No Girls Allowed + The Missing Dinosaur Bone** When Sister beats Brother in running, marbles, and baseball, the boys form a exclusive club. Rather than force them

to admit girls, Mama Bear finds a better way to reunite the sexes. Second story (reviewed previously) has the Bear family tracking down a museum thief.

Black Beauty (Hanna-Barbera, 49 min., ages 6–12) Animation of Anna Sewell's adventure from an equine point of view. Well-condensed to retain its important events and ideas, the story shows how a horse's temperament relates to its treatment by humans. *Beauty* is nicely drawn and animated to build empathy with the horses. A good intro that should inspire reading the classic.

***Charlotte's Web** (Paramount, 94 min., ages 6–10) E. B. White's fable of friendship, dedication, and appreciation of nature's changes is beautifully drawn and animated by Hanna-Barbera. The loving spider Charlotte spins web after web declaring Wilbur the pig to be RADIANT and TERRIFIC. Not only is he spared the axe, but Wilbur becomes famous as SOME PIG. Because Charlotte dies in the end, you should watch with your child, something you'll thoroughly enjoy. The characters are endearing, songs from the Oscar-winning Sherman brothers lively, and animation—especially as Charlotte spins her miracle webs—full of delights. As her words sparkle in their lacy weave, they convey a love of language shared.

Corduroy and Other Bear Stories (Children's Circle, 40 min., ages 3–8) Title tale has a teddy bear trying to replace the missing button from his overalls, so someone will buy him. His mall adventures are funny and narration-free for the very young. In the second story, bear and tiger imagine *Panama* to be the perfect place, if only they could find it. And Robert McCloskey's *Blueberries for Sal* involves the mix-up of Sal and her mom with a little bear and his mom.

Count of Monte Cristo, The (Hanna-Barbera, 47 min., ages 10–14) The Alexandre Dumas tale of injustice avenged is well written and edited, if only passably animated. Wrongly accused of treason, a sailor escapes from prison, finds treasure, and becomes the Count of Monte Cristo. His story takes even more turns from there until the final surprise move. A good intro to the swashbuckler.

Curious George, Volume 1 (SVS, 30 min., ages 2–6) "George was curious." If this phrase brings a smile to your face, you already know the Margret and H. A. Rey books. The little monkey's six stories on this video have the usual troublemaking followed by restitution. *At the Ballet*, George disrupts and then saves a production of *Jack and the Beanstalk.* He leads a parade when he *Walks the Pets.* And he paints his own face onto a vase at an *Art Show.* Based as they are on the beautifully simple books, both the art and storytelling are excellent for the very young.

Cyrano (Hanna-Barbera, 46 min., ages 10–14) Edmond Rostand's play about the famed French soldier-author is voiced by Jose Ferrer and Joan Van Ark in this animated adaptation. Well-drawn and written, it's true to the original, yet accessible to kids. The large-nosed Cyrano courts Lady Roxanne through the simpler-minded Christian. All the humor and poignancy of the situation comes through in this swashbuckling romance. A good intro to a classic.

Doctor De Soto and Other Stories (Children's Circle, 35 min., ages 3–8) The Oscar-nominated title is based on William Steig's Newbery-honored book. Animated by Michael Sporn and set to light waltz music, *De Soto* is a modern-day fable about mice outfoxing a fox. Its attention to detail, artful lines, and humorous touches ("Would it be shabby, wondered the fox, to eat the De Sotos?") hold appeal for adults as well as children. Other stories are *Patrick* by Quentin Blake, *Curious George Rides a Bike,* and *The Hat* by Tomi Ungerer.

Dr. Seuss videos The fabulous Dr. Seuss, aka Ted Geisel, has been adapted with more or less success. Some videos are better animated than others, some are based on better books to begin with. But all are fairly faithful to the originals, so Geisel's unique art and language translate intact. And his wonderful cast of characters—from Horton to the Cat in the Hat, the Grinch to Sam I

Am—should lead young readers back to the books. The best of the bunch . . .

* **Butter Battle Book, The** (Turner/GoodTimes, 30 min., ages 3–14) Ralph Bakshi animation of Ted Geisel's personal favorite, which took a year to write. The nuclear arms race is paralleled in that of the Yooks and the Zooks, who battle over which side of the bread to butter. Their escalating weaponry may be Seussian fantasy, but their destructive capability is all too real. The animation, also one of Geisel's favorites, is superb, and its message well delivered: We have to learn to live together.

Dr. Seuss's ABC (Random House, 30 min., ages 2–6) This slightly animated video opens with such ABC fun as "Oscar's only ostrich oiled orange owl." Next story has the Cat in the Hat boasting *I Can Read With My Eyes Shut.* And *Mr. Brown Can Moo—Can You?* gets kids mooing and buzzing and slurping in an interactive challenge. Drawing, coloration, and language are all characteristically original in this subtly educational video.

* **Dr. Seuss: The Cat in the Hat** (CBS-Fox, 30 min., ages 3–8) Dr. Seuss's trademark character turns the house of two children into an inside-out, upside-down, roundabout mess. From games like Up Up With a Fish to the kite-flying Things One and Two, the cat's disregard for normalcy is as refreshing as Geisel's poetry. This, as well as his original art, translate well to animation.

Dr. Seuss: The Cat in the Hat Comes Back (Random House, 30 min., ages 3–8) Dissolve animation, where one image dissolves to another to suggest movement, adapts three books on this video. In the title sequel, trying to remove a pink bathtub ring sets off a string of rhyming mishaps. The Cat's so-called helpers this time are little cats A, B, C . . . to Z. *Fox in Socks* pairs up rhymes in the most unlikely ways. And *There's a Wocket in My Pocket* plays with words as an imaginative child might. Seussian nonsense at its best.

Dr. Seuss: Hop on Pop (Random House, 30 min., ages 3–6) This "Simplest Seuss for Youngest Use" adapts three stories with simple animation. Title story is great for young readers, with its minimalist poems of pups in cups and apes gobbling grapes. *Oh Say Can You Say?* twists the tongue with its bed spreaders and bread spreaders. And *Marvin K. Mooney, Will You Please Go Now!* can leave on a zike bike if he likes. Great language play/education.

* **Dr. Seuss: Horton Hears a Who** (MGM-UA, 26 min., ages 3—14) Dr. Seuss's finest allegory has animation to match from Chuck Jones. Because of his large ears, Horton the Elephant hears the tiny voice of a Who in a speck of dust. Each has a hard time convincing their world that the other world exists. This message of open-mindedness and respect shines through a story of such elegance it's as poetic as its language. "A person's a person no matter how small."

Dr. Seuss: How the Grinch Stole Christmas —see Holidays: Christmas

Dr. Seuss: I Am *Not* Going to Get Up Today (Random House, 20 min., ages 2–6) Four books brought to life with "animatics" of camera and cutout motion. Title story tells of a boy whose day it is for "woozy snoozing." Others are *The Shape of Me and Other Stuff*, *In a People House*, and as counterpoint for the opener, *Great Day for Up*. Each offers something unique—poetry, shapes learning, word reinforcement, and all kinds of "ups."

Dr. Seuss: The Lorax —see Family: The Environment

Dr. Seuss: One Fish, Two Fish, Red Fish, Blue Fish (Random House, 30 min., ages 2–6) "Funny things are everywhere" in the title adaptation, from Ned in his little bed to the Nook with his hook and cookbook. Second adaptation celebrates the unbridled imagination is *Oh, the Thinks You Can Think,* like snoves in their gloves. And *The Foot Book* shows feet every which way—high and low, slow and quick, up and down, and clown feet. Only slightly animated, the video succeeds on the strength and humor of the books.

* **Dr. Seuss On the Loose** (CBS-Fox, 30 min., ages 3–10) Three animated fables on the folly of pride and prejudice. *The Sneetches* discriminate against starless bellies until a Star-on ma-

chine confuses the issue. In *The Zax,* two stubborn fools refuse to go around each other as the world grows around them. And then it's the delightful *Green Eggs and Ham,* where Sam I Am tempts his friend into trying the colorful concoction. *Seuss On the Loose* is Geisel at his wisest.

Horton Hears a Who (Random House, 30 min., ages 3–14) Dustin Hoffman *is* Horton. Random House couldn't have found a better reader for this heartwarming fable of vision, tolerance, and caring for all, great and small. Camera dissolves and other movements cleverly animate the book's original art. So the effect is more like being read to, when compared to the musical version. Also included is Dr. Suess's *Thidwick the Big-Hearted Moose*, whose consideration for others has comical results.

Edgar Allan Poe's The Gold Bug (New World, 45 min., ages 7–12) Emmy Award-winning adaptation starring a young Anthony Michael Hall. Set in 1866 on a Carolina island, *The Gold Bug* is a tale of buried treasure with a pirate's curse. The mystery begins when Hall finds a gold bug which leads him to a treasure map written in invisible ink. The adventure afoot, he solves several little mysteries on the way to finding the treasure. Well acted and edited to build suspense to the final moments.

Elephant's Child, The (Random House, 30 min., ages 4–8) A Rabbit Ears Storybook Classic read by Jack Nicholson. Predictably apropos for the snake and crocodile parts, his narration is unexpectedly innocent as the little elephant. Kipling's playful language is in good hands in this tale of how the elephant got his trunk. Also charming are the pastel drawings of the "great, grey-green, greasy Limpopo River."

*__Elizabeth and Larry/Bill and Pete__ (MCA, 25 min., ages 4–9) From Shelley Duvall's Bedtime Stories, *Elizabeth and Larry* is one wonderful tale. Read by Jean Stapleton and based on the book by Marilyn Sadler with illustrations by Roger Bollen, it tells of an oldish woman and her alligator who is more friend than pet. All about the little things friends share as well as deep affection, *Elizabeth and Larry* touches the heart even as it tickles the funny bone. *Bill and Pete*, a Dudley Moore reading of the Tomie de Paola book, also celebrates friendship—this time between a crocodile and his "toothbrush" bird. Painted in the vivid tones of the Nile, *Bill and Pete* shows the success of friends working together.

Encyclopedia Brown: One-Minute Mysteries (Media, 30 min., ages 6–10) Donald Sobol's boy detective solves five mysteries on the likes of bitter drinks and Civil War swords. All involve catching some inconsistency in the story, meaning careful viewing. They're quite fair, however, and solvable to please any kid who loves a mystery.

***Encyclopedia Brown: The Case of the Missing Time Capsule** (Media, 50 min., ages 6–10) Fine adaptation that retains the solid detection of Donald Sobol's book and adds the zest of director Savage Steve Holland. In the mystery of who stole Idaville's time capsule, E. B. and Sally find red herrings among the clues. And they solve several mini-mysteries on the way to an exciting finish. All are well and fairly presented for young crime crackers.

***Five Lionni Classics** (Random House, 30 min., ages 2–6) Contemporary fables from author-illustrator Leo Lionni, animated by Guilio Gianini. Each makes a point of daring to be different, celebrated in Lionni's distinctive and humorous art. There's Cornelius the crocodile who walks upright and Swimmy the black goldfish. Best of all is Frederick the artist, whose gathering of colors and words proves as important as food for the long and dreary winter. A delight to be enjoyed on many levels.

***Five Stories for the Very Young** (Children's Circle, 33 min., ages 2–6) *Changes Changes* animates wooden dolls and blocks in ways to spark young imaginations. Its crescent moon ending segues nicely into the moon of *Harold's Fairy Tale.* With his magic crayon, Harold can live in his drawings. His triangle can be a castle tower or witch's hat or whatever he wants it to be. The alliterative and artful *Caps for Sale* is full of monkey business, as a peddler tries to get his hats back. Rounding out the tape are *Drummer Hoff* and *Whistle for Willie.*

***Fun in a Box: Ben's Dream and Other Stories** (Made-to-Order, 30 min., ages 6–10) Title story tells of a boy who travels the world afloat in his house. Its animated transitions are masterful as Ben drifts from Big Ben to the Eiffel Tower all the way to the Great Wall of China. The finale recaps where in the world he's been via such landmarks. Next up is *Your Feet's Too Big,* animating the Fats Waller number with a dancing menagerie. Finally, *Fish* is a live-action detective story that kids find as funny as it is intriguing.

Fun in a Box: New Friends and Other Stories (Made-to-Order, 30 min., ages 6–10) Title story is an award-winning adaptation of *Howard* by James Stevenson. Howard, a hapless duck who's missed his flight south, finds new friends during his winter in New York City. Excellent animation brings Stevenson's fine drawings to life. Other shorts feature fascinating kinetic sculpture, animated lights fantastic, and storytelling about *Why Cats Eat First.*

Gulliver's Travels (Hanna-Barbera, 60 min., ages 7–14) Jonathan Swift's brilliant satire translates well in this animation. While the emphasis is on its adventure aspects, Gulliver still skewers pomposity, provinciality, and pettiness in his travels. Even the animation is pointed, as Gulliver himself is drawn naturistically compared to the cartoony Lilliputians. Well-written, funny and wise, the video's a fine introduction to the classic.

Happy Birthday Moon and Other Stories for Young Children (Children's Circle, 28 min., ages 2–6) Title story is the Frank Asch charmer about a bear's gift to the moon. *Peter's Chair* features the collage art of Ezra Jack Keats in a tale of new-baby jealousy. *The Napping House* is a lullaby of a story with delightful repetition à la *The House That Jack Built. Three Little Pigs* has appropriately medieval art and juicy narration for a tale where two pigs and the wolf get eaten. And *The Owl and the*

Pussycat rounds out the tape most lyrically. All are beautifully drawn and narrated, if only slightly animated.

Heidi's Song (Hanna-Barbera, 85 min., ages 4–10) Animated musical of Johanna Spyri's classic featuring the voices of Lorne Greene and Sammy Davis, Jr. Whether in her grandfather's Alpine home or her uncle's dreary mansion, Heidi makes the best of a situation. So does Hanna-Barbera in this bright production. Well-drawn, animated, written, and scored, *Heidi* is upbeat without the sentimentality that often mars its adaptation.

Hiawatha (FHE, 51 min., 7–12) Australian animation of the Longfellow legend is well drawn and adapted for young viewers. Opening with his boyhood friendship with Minnehaha, the story centers on Hiawatha's adventure to save his people from famine. He braves giant sturgeon, crayfish, and serpents on his way to meeting the Great Bear. Deeming battle useless, Hiawatha uses a trick learned from Minnehaha to garner the Bear's magic corn seeds. A good adventure well told.

*__How the Leopard Got His Spots__ (SVS, 30 min., ages 4–10) Perfectly conceived, this video presents the Rudyard Kipling Just So Story with fine narration by Danny Glover, original music by Ladysmith Black Mambazo, and the vibrant art of Lori Lohstoeter. Her Fauve-like paintings capture the colors and patterns of Africa. Director Tim Raglin combines all these facets into a jewel of a family video.

How to Eat Fried Worms (CBS-Fox, 25 min., ages 7–12) CBS Storybreak Video based on a book by Thomas Rockwell, son of the famed illustrator. Billy bets he can eat a worm a day for fifteen days, leading to creative culinary approaches. When it looks like he might succeed, his friends try to thwart him. While only passably animated, the story is funny, suspenseful, and full of clever tricks.

I'm Not Oscar's Friend Anymore and Other Stories (Golden, 30 min., ages 4–9) Four book-based animations built on themes of friendship. Boy in the title story predicts all the terrible things that will happen to Oscar after their argument. Creatively animated, it ends naturally in reunion. *Creole* is the neon-bright tale of an "ugly" bird who has to convince other animals that she's not mean. The leanly-drawn and funnily-animated *Hug Me* stars a porcupine who goes to great lengths to get a hug. And *Birds of a Feather* is wordless animation of a bird who chirps to a different drummer.

Joey Runs Away and Other Stories (Children's Circle, 30 min., ages 3–8) Jack Kent's title story is animated to bring out all its humor. Refusing to clean his "room," Joey the kangaroo leaves his mother's pouch. His trials, as well as those of his would-be replacements, are truly funny. So, too, is *The Bear and the Fly*, as one tussles with the other in Paula Winter's story. *The Most Wonderful Egg in the World* animates Helme Heine's tale of three hens in friendly competition. Peter Spier's *The Cow Who Fell in the Canal* rounds out the tape with an adventure-seeking cow. A fine and funny collection for young children.

Jungle Book, The (Walt Disney, 78 min., ages 3–10) Animation of Rudyard Kipling's Mowgli stories with Disney touches of humor, pathos, and song. With the return of Shere Khan the tiger, the panther Bagheera tries to return Mowgli to humans. Easy-going Baloo the bear, however, wants to keep his young friend in the jungle. With its childhood fantasies of living in the wild and befriending animals, *Jungle Book* appeals to a wide age range, right up to adults.

Last of the Mohicans, The (Hanna-Barbera, 50 min., ages 8–14) Animation of James Fenimore Cooper's classic adventure. Escorting two sisters to their father, a major enlists the aid of hunter Hawkeye and last-of-the-Mohicans Uncas against the kidnapping threat of enemy Hurons. The plot moves with captures and rescues, ambushes and escapes. Animation is nothing special, but drawing and storytelling are excellent in creating genuine drama. A fine adaptation for children.

Lion, the Witch, and the Wardrobe, The (Vestron/FHE, 95 min., ages 6–10) This first book of C. S. Lewis's *The Chronicles of Narnia* is animated by Bill Melendez with the Children's Television Workshop. Four children fulfill a prophecy to replace the witch-queen on the throne of Narnia. They enter this fantasyland through a wardrobe and find invaluable aid in the magical lion Aslan, hence the title. Art and animation are just passable, but the eminently childlike fantasy is well told. And its musical score adds considerably to the magic and mystery of it all.

***Little Toot/Choo Choo** (MCA, 25 min., ages 2–6) This Shelley Duvall Bedtime Story opens with *Little Toot and the Loch Ness Monster,* based on Hardie Gramatky's book and read in a fatherly style by Rick Moranis. It tells of a little tugboat's trip to Scotland, where he stands up to bully boats and plays with Nessie. The music is a Scottish delight, and art watercolored with such immediacy you can almost smell the paint. Based on the Virginia Lee Burton book, *Choo Choo* is nicely narrated by Bonnie Raitt. A simple tale of a runaway steam engine, it's beautifully drawn and colored in charcoal with pastels.

Lyle, Lyle Crocodile: The Musical (Media, 25 min., ages 3–9) Tony Award-winning Charles Strouse scored this lively adaptation of *The House on 88th Street.* Tony Randall narrates the story of a family whose new house comes with a crocodile—the caviar-eating, hula-hooping, song-and-dancer Lyle. Michael Sporn animation holds its usual delights while remaining faithful to the book's art.

***Madeline** (Media, 30 min., ages 3–8) Sparkling musical adaptation of Ludwig Bemelmans's classic where the French schoolgirl has her appendix out. Narrated by Christopher Plummer, the story is treated to a fine score by Sesame Street's Joe Raposo. And the animation is a delight, as clever as Madeline herself. Altogether a wonderful complement to the book.

***Madeline and the Bad Hat** (Golden, 25 min., ages 3–8) Pepito, the boy next door, is a bad hat—naughty and haughty and stirring up trouble. Realizing he's probably just lonely, Madeline befriends Pepito who, in turn, befriends the animals he formerly tormented. A song and dance at the zoo makes a fun and fitting finale. With great music and art based on the book, this video really stars the clever rhymes of Ludwig Bemelmans.

Madeline's Christmas —see Holidays: Christmas

***Madeline's Rescue** (Golden, 25 min., ages 3–7) Once again Christopher Plummer reads a musical *Madeline* story. The French schoolgirls adopt the dog who's just rescued their youngest from the river. But with "no dogs allowed" at school, can they keep her? Painted in warm, European colors, the artwork is as charming as that of the book. Add Bemelmans's gift for poetry and the girls' harmonizing in song, and you have another fine *Madeline.*

***Madeline's Rescue and Other Stories About Madeline** (Children's Circle, 23 min., ages 3–7) With characteristic care, Children's Circle adapts three *Madeline* stories: the title book, the *Bad Hat,* and the *Gypsies.* Their faithful adaptation allows the French heroine's qualities to shine. Every bit of her charm—in art, rhyme, and characterization—comes through in these lightly animated stories.

***Maurice Sendak Library, The** (Children's Circle, 30 min., ages 3–8) Animated treasury of the award-winning author-illustrator's work. Stand-out title is *Where the Wild Things Are*, with animation as poetic as the book itself. *In the Night Kitchen* serves up Mickey Cake, with Ollie Hardy chefs and delightful period graphics and music. Animated songs from *Maurice Sendak's Really Rosie* and a short documentary on Sendak round out the tape.

***Merlin and the Dragons** (Lightyear Entertainment, 25 min., ages 5–10) Based on a Jane Yolen story, this animated video is read with dramatic flair by Kevin Kline. Merlin reassures the newly-crowned Arthur with a story from his own childhood. Uniquely gifted, the young Merlin is at first mistrusted but ultimately heralded as his village's savior. With this message of youthful self-confidence, exquisite art and animation, and aura of Arthurian magic, *Merlin* makes great family entertainment.

Mike Mulligan and His Steam Shovel (Golden, 25 min., ages 3–7) Virginia Lee Burton's classic becomes an animated musical in the hands of Michael Sporn. While the songs are little more than filler, the animation is beautifully drawn and colored to bring the book to life. With the advent of higher-powered machines, Mike and his steam shovel Mary Anne leave the city for the small town of Popperville. There they land a job and a future, thanks to the quick thinking of a boy. Fans of the book will not be disappointed.

Moby Dick (MGM-UA, 52 min., ages 8–14) Herman Melville's novel of obsession is surprisingly well adapted in this animation. Driven mad after forty years a-whaling, Captain Ahab pursues Moby Dick against all reason and plea. The young Ishmael narrates the story of the great hunt in all its drama. While stiffly animated, the video is well drawn, written, and paced to elicit the story's adventure. As such, it makes a good children's intro to the classic.

Mouse and the Motorcycle, The (Strand, 42 min., ages 4–10) Beverly Cleary's book is brought to life with a combination of live-action and stop-action animation. New to town and living in a hotel, young Keith befriends Ralph, the mouse he catches riding his toy motorcycle. Seeing life from the mice point of view is as much fun as Ralph's cycle escapades. And animating a realistic mouse makes the fantasy believable. See also *Runaway Ralph,* the sequel starring Fred Savage.

Nancy Drew: The Mystery of Pirate's Cove (MCA, 47 min., ages 7–12) One of eight episodes of the 1977 television series starring Pamela Sue Martin. More entertaining than its Hardy Boys counterpart, *Nancy Drew* shows the young detective at work, observing and deducing until the crime is solved. Nancy's courage and intelligence in action make her a fine role model for girls. Here the mystery is a lighthouse ghost whose existence is "proved" by a parapsychologist. Loaded with atmosphere, the story builds suspense right up to a surprise twist at the end.

Norman the Doorman and Other Stories (Children's Circle, 30 min., ages 4–9) Norman's a mouse who enters a contest staged by the art museum that's his home. Afraid to sign the work and risk discovery, Norman is happily surprised at the end. Also included is Robert McCloskey's *Lentil.* The beautifully-animated *Brave Irene* rounds out the tape. Entrusted with delivering the duchess's gown, Irene braves a terrible snowstorm. She's taunted by the wind and hampered by a twisted ankle but perseveres to the palace. A real charmer.

*__Owl Moon and Other Stories__ (Children's Circle, 30 min., ages 4–9) Title short, which is iconographic rather than animated, tells the quiet story of a boy who goes owling with his father. Also filmed directly from the book is *Time of Wonder,* Robert McCloskey's Maine paean. The wittily-animated *Caterpillar and Polliwog* has two creatures looking forward to a change. Also fully animated is *Hot Hippo,* an African folk tale alive with bright, translucent watercolors. Each celebrates animals for the special creatures they are.

Paddington Bear, Volume 1-6 (Walt Disney, 50 min. each, ages 3–7) Absolutely charming adaptions of the Michael Bond books starring the stuffed toy amid cartoon people. Volume One offers a dozen of Paddington's first adventures, beginning with his being found by a family at the railway station whose name he bears. Whether packing bacon "for emergencies" or playing with shaving cream, Paddington Bear exhibits the sort of light British humor that has endeared him to young children everywhere. On video his simple stories are suitably animated and paced.

*__Pigs' Wedding and Other Stories, The__ (Children's Circle, 39 min., ages 3–10) Porker and Curlytail are in love. Naturally, marriage follows, calling for a most *un*natural cleanliness on the part of their friends. Helme Heine's story is full of delights—in art, animation, music, and humor. Also included is the mesmerizing Celtic tale of *The Selkie Girl*, a sea maiden who comes to live on land. Adding the proper lilt are Jenny Agutter's narration and Ernest Troost's score.

***Pride and Prejudice** (CBS-Fox, 3 hrs. 45 min., ages 11 and up) Jane Austen's finest moment is brilliantly dramatized in this 1983 BBC production. Cyril Coke directs with masterful timing and style. As Elizabeth Bennet, Elizabeth Garvie is simply wonderful. And David Rintoul makes the perfect Darcy. Their verbal sparrings ignite not only their passion but our interest in their comedy of manners. A timeless joy to share episodically with older children.

Prince and the Pauper, The (Walt Disney, 24 min., ages 4–9) In his first theatrical 'toon after a decade hiatus, Mickey Mouse shines in the dual title role. Donald Duck, Goofy, and Pluto play the pauper's friends in the Mark Twain swashbuckler. And true to the story, after trading places due to mutual envy, the prince and the pauper realize each life has its hardships. Smoothly animated with Disney touches of humor, this video offers a bright and bonny adventure.

Reading Rainbow videos When this PBS series encouraging reading first aired, children's librarians were unprepared for the deluge of requests for its featured books. And many credit an increase in reading directly to the program. Host LeVar Burton visits sites relating to the videos' stories, which slightly animate original art. Each video is an engrossing blend of fact and fiction, live-action and animation, and adult and kid viewpoints.

Arthur's Eyes (Vestron/FHE, 28 min., ages 5–10) Builds on the theme of How We Look at Things. Bill Cosby reads the title story of adjusting to eyeglasses. *The Turnabout, Lookabout, Thinkabout Book* is cleverly introduced in a split-screen with four LeVar Burtons arguing about what a picture represents. It all depends on how you look at it!

* **Digging Up Dinosaurs** (Vestron/FHE, 30 min., ages 4–10) Title book is a real charmer, following a girl's trip to the museum where she learns how fossils formed and how paleontologists work. Follow-up includes Burton's visit to Dinosaur National Monument in Utah where such fossils are preserved on site. With animation of a stand-up dino comedian and several book recommendations, *Digging* yields real treasure for the dino fan.

Dive to the Coral Reef + The Magic School Bus: Inside the Earth (Pacific Arts, 60 min., ages 5–10) Burton scuba dives to an underwater park in *Reef*, where he learns about coral's uniqueness as part-animal and part-plant. He does an excellent job, too, of reporting the sights and sensations of the experience. *Bus* features animation of the popular Scholastic book's fantastic voyage to the center of the Earth. Burton goes in as far as *he* can by spelunking with a scientist who shares her expertise on cave formations.

Gregory the Terrible Eater + Gila Monsters Meet You at the Airport (Vestron/FHE, 60 min., ages 5–10) First story tells of a kid goat who, uncharacteristically, is a fussy eater. In a visit to a zoo kitchen, Burton learns what goats really eat, and then kids imagine the wildest foods, like fly salad. The second book adaptation plays on kids' misconceptions about places they're moving to. A New York boy learns Arizona isn't *that* weird in this artfully drawn and funny animation. After talking about his own childhood move, Burton meets a real gila monster.

* **Mummies Made in Egypt + Bringing the Rain to Kapiti Plain** (Pacific Arts, 60 min., ages 6–10) First episode looks at how and why mummies were made, as well as how archaeologists preserve and reconstruct the ancient Egyptians. Animation of the title book fascinates with a detailed presentation of mummification, including the removal of brain tissue through the nose. *Kapiti* celebrates the importance of rain in a cumulative poem à la The House That Jack Built. After hearing kids' theories of thunder and lightning, Burton visits a thunderstorm lab in Colorado to get the real scoop.

Perfect the Pig + Ty's One-Man Band (Vestron/FHE, 60 min., ages 5–10) First story, narrated by James Coco, tells of a winged pig and his human friend Olive. The pig theme continues with visits to a hog farm, a Hawaiian zoo store, and a library where Kermit the Frog waxes poetic on his favorite porker. Music stars in the second tale, narrated by Lou Rawls and beautifully illustrated by Margot Tomes. Young Ty is stunned by the music his park friend can make with the likes of a comb, wooden spoons, washboard, and pail.

Richard Scarry's Best ABC Video Ever! (Random House, 30 min., ages 2–6) Busytown comes alive in this musical video animating Huckle Cat, Lowly Worm, Bananas Gorilla, and other cute Scarry characters. The setting is the classroom and emphasis on fun as they open with the alphabet song. Then each of twenty-six students tells a story focusing on a letter. The airport has an air cargo of apples, while the baker eats bread and bananas for breakfast. An engaging introduction to the letters of the alphabet. Also worthwhile is *Richard Scarry's Best Counting Video Ever.*

*__Rikki-Tikki-Tavi__ (FHE, 30 min., ages 4–12) Orson Welles narrates this excellent animation from Chuck Jones. Rudyard Kipling's story centers on the mongoose with the onomatopoetic name. Bright-eyed and bushy-tailed, Rikki finds his perfect foil in the conspiring cobras, hissing and whispering in the shadows. As a pet to a human family, his courage and curiosity serve them well. Rikki's story moves as briskly as he does, making this an entertaining video for all ages.

Robert McCloskey Library, The (Children's Circle, 55 min., ages 3–7) Adaptations include *Lentil, Make Way for Ducklings, Blueberries for Sal, Time of Wonder,* and *Burt Dow: Deep-Water Man.* Plus there's a profile of the award-winning author-illustrator in which he talks about the importance of seeing and thinking as well as the hands in drawing. The only animated story is also the best. A contemporary Jonah and the Whale, *Burt Dow* is as colorful and amusing as the character himself.

Runaway Ralph (Strand, 42 min., ages 4–9) This sequel to *The Mouse and the Motorcycle* by Beverly Cleary has Ralph leaving home for the wilds of a summer camp. There he befriends a lonely Fred Savage and scuffles with a tomcat. Scooting around on his cycle, Ralph is as cute as ever, especially when besting the cat. Their cat-and-mouse play involves the kind of slapstick and trickery kids love. Runaway fun.

Shelley Duvall's Bedtime Stories The queen of children's programming does it again. She has taken her favorite storybooks, treated them to faithful animation, and found just the right celebrity readers to bring them to life. Duvall herself introduces each story and wraps up the tape with reminders of the joys of imagination. Reviewed in this section are two videos, alphabetized by title, each including two stories: *Elizabeth and Larry/Bill and Pete* and *Little Toot/Choo Choo.*

Sherlock Holmes and the Baskerville Curse (Strand, 67 min., ages 8–12) National Education Association-recommended animation with Peter O'Toole supplying Holmes's voice. While the animation is rather stiff, its storytelling is sound. Sir Arthur Conan Doyle's most famous work is treated to an atmospheric and dramatic retelling. And it's excellently drawn, with naturalistic characters and a stunning phosphorescent hound. Keeps you hooked even when you know the story.

Sherlock Holmes: The Priory School (MPI, 50 min., ages 11 and up) One of the more kid-accessible mysteries from the Granada Television series starring Jeremy Brett. His definitive portrayal—of a Holmes in command; curt, concentrated, and cuttingly humorous—is at least half of the show's entertainment. The other half, of course, is following his detection of a seemingly impossible crime. Here the son of the duke is abducted from his school, leaving only confusing tracks. Ending in a torchlit cavern, *Priory School* keeps you hooked from start to finish. Dozens of these Sherlock Holmes titles are available from MPI.

***Snowman, The** (Children's Circle, 30 min., ages 4–9) Academy Award-nominee based on the Raymond Briggs book. Fully animated and without narration, the video is as silent and beautiful as snowfall. A boy and his snowman share the wonders of each other's world in scenes that positively glow. Frame by frame, a work of art to be treasured by the whole family.

***Stories from the Black Tradition** (Children's Circle, 52 min., ages 5–14) And what a rich tradition it is, in five stories based on children's literature. *A Story A Story* animates the Gail E. Haley book of the Anansi legend. A spider-man who cleverly tricks the sky god out of his stories, Anansi has a tale rife with the colors and patterns of African textiles. *Mufaro's Beautiful Daughters* vie to be the king's wife in this tale featuring gorgeous, detailed artwork. Unique cloisonné-like drawings enhance the funny fable, *Why Mosquitoes Buzz in People's Ears.* Rounding out the tape are *The Village of Round and Square Houses,* about life in West Africa, and the sweet, childlike *Goggles!* by Ezra Jack Keats.

Story of 15 Boys, The (Celebrity, 80 min., ages 6–10) Animation that adapts the only novel Jules Verne wrote for children. Set in 1860, students of the Chairman's Boys School shipwreck and must pull together for survival on a deserted island. Along the way they encounter bats, skeletons, and yes, pirates. While only passably animated, *15 Boys* is an excellent and well-drawn adventure.

Tales of Beatrix Potter, The (Vestron/FHE, 46 min., ages 2–6) Miss Potter's original drawings are, fittingly, only slightly animated in this video. Their inherent charm, plus fine music and narration, bring over a dozen stories and classic rhymes to life. Most enjoyable are *Peter Rabbit, The Tale of Two Bad Mice,* and *The Story of Miss Moppet.* All show such a love of language, they could spark a lifelong affair in your little one.

Thomas the Tank Engine and Friends videos (Strand, 40 min. each, ages 3–7) Picture a train set driven by a child's imagination and you have the idea of Thomas the Tank Engine. Introduced to American audiences on the Emmy Award-winning "Shining Time Station," *Thomas* is based on The Railway Series by the Rev.W. Audry. Ringo Starr makes the perfect narrator, giving life to all the engine personalities, both cheery and cheeky. Animated by stop-action, their clashes and resolutions shape seven stories on each tape. The stories offer a world of their own and show positive values in action—mainly respect, empathy, and getting along.

Better Late Than Never In the title story, Thomas learns that sometimes late is better, and so he shouldn't tease the slower Henry. Other stories tackle pride, bad manners, and jumping to conclusions. With titles like *Pop Goes the Diesel* and *Gordon Takes a Dip*, their lessons are cleverly conveyed in little dramas and comedies.

* **James Learns a Lesson** More tales of wisdom and folly at Sir Topham Hat's railyard. The first three tell of James's adjustment to pulling coaches after hauling freight. Thomas learns a lesson, too, when he takes off without a conductor. And in *Thomas Goes Fishing*, his trip on the branch line means getting water from the river, with amusing results. A delight to watch, with English countryside sets and amazingly expressive engine faces.

Tenders and Turntables Title story finds the tender engines on strike over a malfunctioning turntable. *Thomas Comes to Breakfast* is the tongue-in-cheek title for this tale of conceit that ends in his crashing into the stationmaster's home. And when Henry pulls *The Flying Kipper* filled with fish from the harbor, he runs into snowy trouble. A nice balance of exciting and humorous adventures.

* **Thomas Breaks the Rules** More goings-on at Sir Topham Hat's railway—all the teasing and boasting that precede amiable endings. In the title story, Thomas is ticketed for failing to have a cowcatcher and sideplates. Other stories have engines foolishly bragging about their color, speed, or special talent. All end in the realization that everyone has flaws and deserves respect anyway.

* **Thomas Gets Bumped** Best of these timeless tales of rivalry and friendship have the vehicles preparing for a children's garden party, dealing with a visiting "celebrity" engine, and playing little practical jokes on each other. In his second video, George Carlin again makes a fine and funny narrator. And the sets, reminiscent of old-fashioned toys, are more charming than ever.

* **Thomas Gets Tricked** Title story introduces Thomas as a "cheeky little engine" who taunts the bigger ones as "lazy-

bones." Naturally, the tables turn in another story, so each learns to appreciate the skills of the other. Remaining stories stress co-operation, patience, and learning before doing. The engines' feelings during all this—sad, happy, fearful, angry—are charmingly expressed on their round, round faces.

* **Trust Thomas** The first video where George Carlin ably fills the talented shoes of Ringo Starr. Rather than feign a British accent, Carlin exploits his own talents in characterization and storytelling. The stories he tells revolve around forgiveness, bragging, keeping promises, accepting advice, and, of course, trust. One especially nice tale has the engines and cars cooperating to replant the storm-ravaged *Henry's Forest.*

***Velveteen Rabbit, The** (Random House, 30 min., ages 3–8) Margery Williams's tale of magic and love finds poetic expression in this Rabbit Ears Storybook Classic. Meryl Streep makes a wonderful reader, and David Jorgensen's pastel art is velveteen soft and warm. Just as the rabbit becomes real when loved by a child for a long time, this video should become a classic the same way.

White Seal, The (FHE, 30 min., ages 6–10) Rudyard Kipling's *Jungle Book* story is treated to fine animation by Chuck Jones and narration by Roddy McDowall. Fascinating scenes of sea life precede the education of the White Seal to protect him from men's threat. Shown as a dark menace with clubs, the seal hunters may frighten young children. Older kids, however, should enjoy the White Seal's quest for an island where no men go.

***Winnie the Pooh and the Blustery Day** (Walt Disney, 24 min., ages 3–7) This Oscar-winning adaptation stars a wind so forceful it blows the words right off the pages of Pooh's book. After a dally at his Thoughtful Spot, the little bear's adventures run from Owl's broken house to Tigger's first appearance and Piglet's rescue from a raging flood. Updated a notch for today's children, Disney's treatment retains all the sweet and funny charm of the A. A. Milne classic.

Winnie the Pooh and a Day For Eeyore (Walt Disney, 25 min., ages 3–7) Poor Eeyore—will anyone notice him on his birthday? Not only is it a rainy day, but there's a cloud following him *and* he's been bounced into the river. His friends come to the rescue, of course, with a party and a game of Pooh Sticks. Sweet, whimsical characters in a story to match.

Winnie the Pooh and Tigger, Too (Walt Disney, 25 min., ages 3–7) The Unbouncing of Tigger is the order of the day. And it's accomplished in a way that pays homage to its book origins. Other connections with the book are made—can the narrator talk Tigger down from the treetop?—to encourage reading. One of the best scenes has Pooh, Rabbit, and Piglet haplessly lost in the woods.

***Winnie the Pooh and the Honey Tree** (Walt Disney, 24 min., ages 3–7) Alas, the "bear of very little brain" is out of honey. Neither his "Rumbly Tumbly Song" nor raincloud disguise can garner him the precious gold. So he eats at Rabbit's Howse to the point of getting stuck in the hole on the way out. Full of funny sights and notions, this is classic Winnie the Pooh. It's like spending a summer afternoon in childhood imagination.

Winnie the Pooh, The New Adventures of (Walt Disney, 44 min. each, ages 3–8) Disney series creating new stories for the friends of Hundred Acre Woods. Fortunately, care has been taken to preserve the original's gentle humor and character-driven stories. Typical of its plotlines is Volume 7's *King of the Beasties,* where Tigger is so crowned after tricking others into thinking a jagular is on the loose. With more drawings per minute than usual in TV animation and nearly twice the production time, Pooh's *New Adventures* are as good as the classics.

Winnie the Pooh (CBS-Fox, 57 min., ages 5–9) A new look at the old bear which is truer to the A. A. Milne books than Disney's treatment. Circuitous, confusing conversations and ruminations are well-delivered by storyteller Willie Rushton. Periodically, the original drawings of Ernest H. Shepard punctuate the four tales: *Kanga and Roo Come to the Forest, Pooh*

Invents a New Game, Rabbit Has a Busy Day, and *An Enchanted Place.*

Wizard of Oz, The (Sony Kids Video, 30 min., ages 3–6) Purists may balk at this greatly condensed, simply animated version of L. Frank Baum's classic. Yet it's a nicely drawn and well-written introduction for the very young. Downplayed are the threats of violence and emphasized are friends' cooperation and humor, especially from the Scarecrow. A good preschooler-friendly adventure.

EDUCATIONAL

"Educational TV" conjures up images of stuffy teachers in even stuffier settings lecturing on the stuffiest of subjects. Not so with these videos. Their creators know the value of entertainment in education and have put together quality productions. Learning is fun whether the subject is the alphabet, counting, dinosaurs, or other animals. Video series, such as "Mister Rogers" and "Sesame Street," are listed under those headings for easy location.

Animals Are Beautiful People (Warner, 92 min., ages 4–10) Filmmaker Jamie Uys (*The Gods Must Be Crazy*) pays tribute to the creatures of his native Africa. The best scenes tell stories—funny in the case of playful baboons, sad when baby pelicans are left behind in a drought, and dramatic in a cheetah's chase of a warthog. A film of wonder and natural beauty.

Animals of Africa, Volumes 1-9 (Celebrity, 70 min. each, ages 6–12) Recommended by the National Education Association, this nature series is hosted by Joan Embery of the San Diego Zoo. She's a bright narrator, too, using fine footage of not only animals but also the landscape of Africa. Even the background music is excellent. And the education goes beyond the usual facts to explore survival tactics, such as adaptive coloration. Titles of most interest to kids: *Big Cats of the Kalahari*, *Wondrous Works of Nature*, *The Land of the Elephants*, and *Paths of Survival*, focusing on young animals' learning.

Baby Vision, Volumes 1 and 2 (J2, 40 min. each, ages 2–5) Like New Age video for toddlers, these tapes set simple images to soft, natural music with little narration. A segment on Liquids, for example, pictures waterfalls, faucets, poured milk, and fruit juices. One clever sequence tranforms fresh fruit nestled in a glass pitcher into juice as if by magic. Other segments include Motion and Sound, Plants and Flowers, and Dreams.

Bill Cosby's Picture Pages (Picture Pages, 30 min., ages 3–6) "Picture Pages are the epitome of what I believe in . . . educating children with love and laughter," notes Bill Cosby in a letter to parents that accompanies these videos. An activity booklet also comes with each tape, so your child can do each Picture Page right along with Cos. Like most good teachers, Cosby knows the value of entertainment in education. Not only are there funny skits in-between, he plays around as he works the pages. And their education goes beyond the staples of letters, numbers, and shapes to include lessons in preconceptual thinking, set theory, and relational words. You can help by doing the booklet's follow-up activities with your child. Some of the best of these National Education Association-recommended programs . . .

* **Clear Thinking** Since learning how to approach a problem is as important as solving it, these Picture Pages stress thinking through things. Most involve match-ups—of weather and clothing, for example, or animals and their homes. In others, Cosby shows how to order scenes sequentially, as in pizza-making. After one Picture Page reveals a bucket as not belonging in a set of vehicles, Cos looks on in disbelief as a child rolls by on a bucket with wheels. Such humorous touches celebrate free thinking as well as clear.

Numbers Not only does this video cover the fundamentals of numeral recognition, sequence, and counting, it cleverly introduces a number line. Cosby's filling in its missing numerals is a deceptively simple exercise which lays the foundation for a lifetime of math understanding. Because of the way they're worked, these Picture Pages also reinforce the prereading skill of left-to-right eye movement.

* **Shapes and Colors** This video is chock-full of fun education. The opening Picture Page has black-and-white drawings of fruit on top and circles of color below. As Cosby matches the

fruit to its color, you realize this is no simple exercise. Apples, for example, can be red, yellow, or green. The next Picture Page teaches colors, shapes, sequence, *and* safety in a traffic matching game. In another page, Cos and your child find hidden shapes in a grocery store. Not only can you play this game yourself while shopping, there are other good follow-up activities suggested in the booklet.

What Goes Where In matching games, rebus equations, and connect-the-dots, Cosby teaches children about more than the world around them. He emphasizes taking time and thinking about possible answers as he works each page. Well-paced and peppered with funny sketches from Cos, *Where* has pages of vehicles, sea creatures, and rhyming pictures.

* **What's Different** Also covers what's the same—as in a Picture Page where pairs, such as a rowboat and a motorboat, are each matched to a turtle or a rabbit. Another excellent page teaches relationships such as finger:hand::toe:foot. And it's all done with great humor from Cos who claims to *be* a Picture Page in one skit and dances The Funky Galoshes in another.

Words and Letters Going beyond letter recognition, sequence, and phonics, this video uses rhymes and rebuses to help young readers. One Picture Page has pictures to match by rhyme, such as car and star, hook and book. Another joins pictures to form compounds, such as butter and fly to make butterfly. Fast-paced and assuming some reading skills, *Words* is best for older preschoolers.

***Charles the Clown** (A&M, 30 min., ages 3–8) An unassuming title for a very fine and funny video. Before a young audience, Charles Kraus literally becomes a clown—warming up with a few jokes, applying make-up and costume, and finally performing his act. With deceptive simplicity, Charles entertains his audience, even via video, in this celebration of a classic art form.

Circus ABCs (FHE, 30 min., ages 2–6) As fun as it sounds. Opening with the alphabet song, the live-action tape goes on to show A for Acrobat, B for Big Brown Bear, C for Clown, and so on. Alliteration accentuates the circus acts, as do catchy music and clear narration. And the pace is brisk enough to hold young attention spans and yet slow enough for children to "get it."

Circus 123s (FHE, 30 min., ages 2–6) Same format as *ABCs* but with one ringmaster, two trapeze artists, three bears, etc. The tape goes beyond numeral recognition to cover the different concept of number, counting, review, and then repetition of the whole process with new circus acts. Another bright production with funny clowns and animals to spur the learning.

Clifford's Fun With . . . series (FHE, 30 min., ages 3–7) These Scholastic Learning Library videos star Clifford the Big Red Dog, whose clumsiness adds a touch of humor. With owner Emily Elizabeth, he learns his lessons within stories that are slightly animated and scored with lively music. Best of the bunch are:

* **Clifford's Fun With Numbers** Clifford's Birthday Surprise is the occasion for number learning as he shops for one cake, buys two gifts, rescues three kittens, has four paws, and gets five of everything from the party-supply store. With the likes of a salami-salmon cake and Clifford's tornado of a sneeze, there's a lot of humor baked into this birthday surprise.

Clifford's Fun With Rhymes The Rhyme Cats is the club to join, but to be a member you have to solve a rhymed puzzle. Clifford and Emily Elizabeth hone their rhyming skills in games your child can play, too. What rhymes with nose, they wonder, while circling a house with a prominent hose.

* **Clifford's Fun With Shapes and Colors** The Scavenger Hunt is the setting for learning about shapes and colors. Clifford and Emily Elizabeth search for such things as "a circle you can wear" and "something whose name is a color." With reinforcing songs and pauses to allow your child time to answer, this is truly fun learning.

***Dinosaur!** (Vestron/FHE, 50 min., ages 4–10) Christopher Reeve hosts this classic look at the prehistoric beasts who fire young imaginations. The program shows not only what we know about dinosaurs but how we know it—through the detective work of paleontologists. And it entertains with a well-paced variety of segments, including animation of a duckbill family and of a Tyrannosaurus rex vs. Monoclonius bout, which gets a bit bloody.

Dinosaurs! (Golden, 30 min., ages 4–10) When Fred Savage can't come up with a science report, he dreams of dinosaurs coming to the rescue. After a little research to learn more, he opens his report with animated chalk drawings. Then Will Vinton Claymation takes over—of dinosaurs famous (Supersaurus), infamous (Tyrannosaurus rex), and ordinary (herbivore families). Educational video with the emphasis on entertainment.

***Fraggle Rock, Volumes 1–10** (Walt Disney, 30 min., ages 3–8) A clear high note in Jim Henson's career, *Fraggle Rock* offers a world of its own. The Fraggles fancy themselves the "pinnacle of civilization" who only want to dance and sing all day. Their little neighbors the Doozers are tireless builders, and their huge neighbors the Gorgs the friendliest of adversaries. Colorful and engaging, the five main Fraggles have personalities that are expressed in classic Muppet form. And their stories gently educate how friends should get along. A true family delight.

***Free to Be . . . You and Me** (Vestron/FHE, 50 min., ages 3–9) A classic hosted by Marlo Thomas. The liberation of both sexes is celebrated in music, humor, and animation. Boys learn it's okay to cry and have a doll, girls that they can compete and be more than sweet. One of the best clips is animation of the revisionist fairy tale, *Princess Atalanta,* where she also runs in the race to determine her husband. With a few reminders to moms and dads, *Free* is an entertaining lesson for the whole family.

Here We Go!, Volumes 1–3 (Celebrity, 30–60 min. each, ages 3–8) Live-action series about vehicles narrated by Lynn Redgrave and set to snappy music to keep things moving. Volume 1 shows a helicopter, bulldozer, steam locomotive, Hovercraft, and blimp. Volume 2 an aerial tramway, double-decker bus, ocean liner, fire engine, hydrofoil, and milk truck. And Volume 3 covers over a dozen movers, including a cable car, subway, tractor hayride, airplane, and English punt. Kids themselves are often passengers, and women the vehicles' drivers at times. An engrossing and upbeat look at how things move.

Imagine That! (Golden, 30 min., ages 3–6) Jesse's imagination is so strong, he needs to be drawn into his preschool activities. When he can't draw a real animal in art, though, his imagination saves the day. This video has the real feel of preschool and is an original idea freshly executed. Cute songs and animation make it a good introduction to upcoming preschoolers.

Land of Pleasant Dreams, The (Bridgestone, 30 min., ages 2–6) Series of videos where children's dreams help them solve a problem. Featuring marionettes, each opens with a sewing grandma and moves to a child with his or her stuffed toy in bed. Then it's off to The Land of Pleasant Dreams, with its quilted landscape of Easter basket colors. Volume 3 includes two stories—*A Fence Too High,* about trying until you succeed, and *A Tailor-Made Friendship,* about sticking with a friend in need. Good value education within sweet little stories.

Mine and Yours (Golden, 25 min., ages 3–6) A video potpourri of stories and songs that benefits from Golden's collaboration with Playskool. As expected from these companies, the production is bright and colorful, as Katie struggles with her desire to keep Renee's birthday gift. Through live-action and animated scenes, she learns about sharing and getting along.

Mister Rogers videos (CBS-Fox, 50–60 min., ages 3–7) "Mister Rogers' Neighborhood" is simply wonderful. "Simply" because Fred Rogers distills the richness of life to a child's perspective. And "wonderful" because it works. His award-winning "Neighborhood" is the longest-running children's program on public television. And while Rogers makes it look easy, there's a depth of education in his Neighborhood, not just about things but about feelings and self-esteem. Here are his best home videos.

* **Music and Feelings** Like an extended "Mister Rogers' Neighborhood," this special is a genially-paced montage of guests, visits, and Make-Believe stories. Master cellist Yo-Yo Ma

and the spirited Ella Jenkins share their unique musical expressions. There's a fascinating segment on how bass violins are made. And in The Neighborhood of Make-Believe, King Friday declares a bass violin festival. Since only he can play the instrument, the neighbors have to come up with creative ways to participate, with comedic results.

Talks About Dinosaurs and Monsters In his reassuring way, Rogers stresses that neither can hurt you—dinosaurs because they no longer exist and monsters because they never did. Bad dreams and scary programs are also not real, he adds, and so cannot harm you. In the Neighborhood of Make-Believe, Prince Tuesday is frightened by a monster who turns out to be a costumed Panda sent by the mischievous Lady Elaine. This segues nicely to Rogers himself describing how he used to play monster as a child.

What About Love? Fred Rogers takes on the complexity of love and its array of feelings. When a fight between the King and Queen upsets Lady Aberlin, she asks her friends, What is love? She learns that understanding love is hard because it includes happy, sad, and angry feelings. Fred Rogers works the theme musically with a song and a visit with a percussionist who shows how to express strong emotions creatively.

When Parents Are Away Fred Rogers handles separation anxiety with characteristic care, imagination, and emphasis on feelings. His visit to a day-care home shows some of the activities a child could expect. And in the Neighborhood of Make-Believe, Prince Tuesday faces his first separation from his parents. Both should help pave the way for the first day of school or child care.

*__More Preschool Power!__ (Concept, 30 min., ages 2–6) Like its predecessor, *Preschool Power!*, this tape is simply fun to watch. With original music and a kids-only cast, Montessori methods in such skills as teeth-brushing and funnel-pouring are as entertaining as segments in dancing and silly walks in the park. Funniest clip has a boy searching for flippers in his "pigsty" of a bedroom. Altogether an excellent production.

Mr. Wizard's World: Puzzles, Problems, and Impossibilities (CBS-Fox, 46 min., ages 8–12) Mr. Wizard, aka Don Herbert, has been turning kids on to science since the 50s when his program first aired. By using everyday objects in "magical" experiments, Herbert makes the science as fascinating as the illusion. Even mathematics is shown to hold magic in a series of number tricks. Using fire and chemicals, this video is for older children with adult supervision. No problem—you'll have fun learning, too.

*__National Geographic Video series__ (Vestron/FHE, 60 min. each, ages 6–14) Renowned for its stunning photography, *National Geographic* magazine's move to documentaries was only natural. Not surprisingly, its programs have won hundreds of international media awards. Their education is sound and location footage amazing in view of what the locations are. The sense of being there in the most distant and precipitous situations is broken only by the occasional "How *did* they get these shots?" Some of the best and interesting to kids of fifty *National Geographic Video* titles are:

Australia's Improbable Animals Opening with a geographic explanation of how Australia came to be populated by its unique animals, the video goes on to show marsupials, such as the kangaroo and koala, and other oddities. The wombat and platypus you may know, but how about sky-gliding possums and mudskipper fish who spend more time on land than in water? Great fun for creature lovers.

Invisible World, The Through the specialized "eyes" of cameras, we can see usually hidden worlds. Explored are the wonders of slow-motion photography as well as photo-finish "fast-motion." But the most fascinating worlds are those so small, it takes incredible magnification for us to see—a drop of water teeming with plankton, a dividing cell, and the process of crystallization. The last's breathtaking segments blur the line between science and art.

Lions of the African Night After gorgeous sunset footage of the African bush, the video launches into a typical yet extraordinary night. Shown are the night moves of not just lions, but deer, baboons, hippos, and insects. Shot with low light and vantage point, the effect is one of being among a pride of lions on the prowl. Graphic, and at times bloody in its hunting scenes, *Lions* is best shared with older children.

Search for the Great Apes The real-life *Gorillas in the Mist,* this video tells the story of Dian Fossey and a colleague, Birute Galdikas-Brindamour. The latter is shown trying to introduce a baby orangutan, born in captivity, to the rainforest of Borneo. Fossey, of course, worked in Africa to learn about and protect the mountain gorilla. The best scenes show her beloved Digit as the gentle giant she believed these creatures to be.

Sharks, The Counteracting the hype surrounding *Jaws,* this video claims that humans are more often the remorseless killers and sharks our helpless victims. Substantiating scenes, as well as those of their natural habitat, create real-life drama. One of the best segments shows the detective work of scientists trying to figure why sharks go for their metal cages rather than nearby meat. Be aware that there are some graphic scenes of sharks eating fish whole and of sharks in death throes after being harpooned.

Never Talk to Strangers (Golden, 25 min., ages 3–6) Animated Learn About Living tape that opens with the title music video. Next up is a story, *I Had a Bad Dream,* where a father comforts his son after a nightmare. Best of the bunch is *The House That Had Enough.* Young Anne's messiness and carelessness cause things to run out on her—her bed, clothes, and finally the whole house picks up and leaves. A funny way to make a good point.

Newton's Apple, Volumes 1 + 2 (Pacific Arts, 60 min. each, ages 7–12) Packaging episodes of the award-winning PBS series, these videos truly "make learning science fun." Viewer questions are answered entertainingly with expert visits and clever demonstrations. Volume 1 covers such topics as dinosaurs, bulletproof vests, comets, and perpetual-motion machines. Volume 2 answers questions on skiing, beavers, baseball bats, and polarized light.

*__Peek-a-Boo!__ (Lorimar, 30 min., ages 2–3) Jim Henson creation starring toddlers and Muppets of a teddy bear and bunny rabbit. Short segments entertain with the stuff of young lives—balls and balloons, swings and dress-up, and lots of smiling little faces. A running joke has diffent characters jumping into the screen with a *Peek!* Produced with characteristic care and intelligence, this video makes an excellent intro to the medium. Worth seeking out if not in current release.

*__Preschool Power!__ (Concept, 30 min., ages 3–6) A most original tape, subtitled *Jacket Flips and Other Tips.* Preschoolers themselves demonstrate Montessori methods in over twenty brisk and entertaining scenes. Skills, such as buttoning, pouring, and hand-washing, are clearly shown and set to original music. Wisely interspersed are movement activities to get the wiggles out. See also *More Preschool Power!* and *Preschool Power 3.*

*__Preschool Power 3__ (Concept, 30 min., ages 2–6) Another winner showing kids at play in the most educational ways. Over a dozen clips have preschoolers cleaning, singing, dressing, baking, and making paper fans. In one dreamlike sequence, children play act a fairies' tea party. Another sets up domino chain reactions, including a humdinger that ends in ringing a bell. Wrapping it up is a song called "Together," about cooperating to get things done. Outstanding fun.

Sesame Street videos (Random House, 30 min. except where noted, ages 2–7) (See also Best Series list) The best of the best from the CTW program garnering twenty-eight Emmys and over a hundred international awards in children's programming . . .

Alphabet Game, The It's The Alphabet Treasure Hunt game show with contestants Gary Grouch, Dimples the Dog, and Big Bird. To win, they have to find an item beginning with the letter on the magic game board. Despite trying to lose, the Grouch pulls ahead. And the dog succeeds by pure accident, leaving poor Big Bird the loser. Your child wins, however, by playing along and enjoying musical clips from Sammy the Snake,

Wanda the Witch, and Kermit doing a bayou-themed Alphabet Song.

* **Bedtime Stories and Songs** Wonderful video collecting sleep-themed songs and stories. Cookie Monster sings to a cookie moon, the Count counts sheep, Maria reads a bedtime story, and Ernie dances to sleep with his tap-dancing sheep. All these help Big Bird, and with luck your child, lull his way into dreamland.

Big Bird's Favorite Party Games An interactive video that's like a fun half-hour at a child-care center. Its ten games are designed to get kids up and moving, either to snap and tap a song along or to play Oscar Says. Familiar signing songs include "The Wheels on the Bus" and "In a Cabin in the Wood." Good party fare.

* **Count It Higher: Great Music Videos From Sesame Street** This title is not hyperbole. These *are* great music videos, especially "Doo Wop Hop" and "Letter B" by the mop-topped Beetles. This is the Muppets at their best—hoppin' and boppin' to truly feel-good music. Sesame Street rarely hits a false note, here it hits a high one.

Getting Ready for School As Big Bird flashes back to his first day of school, he remembers being scared but then reassured by seeing on the board "the same old alphabet as on Sesame Street." With his friend Snuffy, he talks about numbers, teachers, and school rules and signs like EXIT and BUS STOP. With Sesame Street's artful blend of live-action and animation, story and song, *Getting Ready* well paves the way to that milestone day.

* **Getting Ready to Read** Not only does this video cleverly reinforce such prereading skills as letter sounds and left-to-right, top-to-bottom eye movement, it motivates the learning with humor. Grover's antics with a WALK sign that actually walks and Cookie Monster's imperfect language help engage children in their struggle to learn. And rhyming games are more than fun—words that sound alike often look alike, and if you can read one, you're well on the way to reading the other.

Sesame Street Special: Big Bird in Japan (60 min.) Some say to get to know a place, you should get lost in it. Big Bird and his dog Barkley unwittingly take this advice and learn about Japan from a friendly woman. After a convertible ride though Tokyo, she takes them to her home, Mount Fuji, and an elementary school. There, first graders perform a lovely fairy tale called *The Bamboo Princess*. It's an entertaining visit that includes learning a few Japanese words.

Sesame Street Special: Christmas Eve on Sesame Street —see Holidays: Christmas

* **Sesame Street Special: Don't Eat the Pictures–Sesame Street at the Metropolitan Museum of Art** (60 min.) Antics of the Sesame Street crew, locked in a museum at night, make a fine intro to the world of art. As different groups search for the lost Big Bird, they encounter art with a child's honesty and humor. Guess which Muppet discovers "Pictures exciting but not for biting"?

Sesame Street Start-to-Read Video: Don't Cry, Big Bird Three stories of accommodation use dissolve animation of the books and are accompanied by text for beginning readers. In the title tale, Big Bird's little friends make their games bigger so he can play along. In *Wait For Me!*, Elmo's grandpa takes his time so the little guy can keep up. And *Grover and the New Kid* learn that sharing toys and waiting your turn work best for everybody. Good values well and simply presented for young children.

Sesame Street Start-to-Read Video: Ernie's Little Lie Being fair is the theme of three stories told with dissolve animation and subtitled text. In the title tale, Ernie submits his cousin's drawing in a contest. *It's Not Fair!* cries Bert when Ernie gets praise for Bert's lemonade. And *Why Are You So Mean to Me?* tells of Grover's attempts to learn baseball. Nicely paced and scored, the stories are also engagingly read by Caroll Spinney, aka Big Bird.

Sesame Street Visits the Hospital Maria plays parent to Big Bird, whose sudden illness sends him to the hospital. There he encounters people and situations unfamiliar but carefully explained along the way. In addition to information about a hospital stay, *Visits* has plenty of humor, music, and talk about feelings. It's very honest in this, too, as Big Bird gets cranky and cries when he hurts. Along with *Visits the Firehouse,* this Sesame Street video shows that good documentaries can be made for preschoolers.

Shamu and You, Volume 1: Exploring the World of Mammals (Video Teasures, 30 min., ages 4–10) An entertaining look at animals, combining live footage, animation, songs, and kids' imaginings. Special spotlights are thrown on elephants, whales, and monkeys, with other segments showing mammal oddities and champions, i.e., largest, fastest, etc. The kids, cute and natural as they talk about their favorite animals, emerge as the best mammals on the video. Other volumes look at fish, birds, and reptiles.

*__Son of Dinosaurs__ (MPI, 60 min., ages 4–10) From the creators of *The World's Greatest Dinosaur Video* comes this sequel packed with more dino lore. When Gary Owens and Eric Boardman receive a live dinosaur egg, they decide to do another show to learn about caring for it. Their visits include Knott's Berry Farm's Kingdom of the Dinosaurs, Alberta's Dinosaur Provincial Park, and interviews with paleontologists on the latest theories. Grand finale is animation of the baby dinosaur hatching from its 120-million-year-old egg. *Son* is a worthy successor.

Tell Me Why, Volumes 1–18 (Prism, 30 min. each, ages 7–12) Based on the Arkady Leokum books, these videos also answer kids' questions factually with a bit of added fun. They're jam-packed with information and have a quick search system to find your way through it all. Eighteen encyclopedic volumes use live-action and animation to educate in such areas as flora, fauna, geology, space, physics, sports, medicine, and government.

Where in the World series (Encounter, 30 min. each, ages 7–12) This Emmy Award-winning program calls itself a geography series but also covers a place's history, culture, and ecology. Each visit combines location footage, colorful graphics, and original music to show how people live around the world. With strong production values, *Where in the World* videos are truly both educational and entertaining.

Kids Explore Alaska The club's Alaskan project opens with Dan's desire to make a totem pole. His visit to a Native American woodcarver shows how and why they're made. Other segments cover Eskimos, gold, wildlife, bush pilots, and dogsled races. Well-paced presentation shows our largest state to be a wilderness worth preserving.

Kids Explore America's National Parks While showing scenes from many parks, this video highlights the natural treasures of Yosemite, Fort Vancouver, and Olympic National Parks. The acting and music may not be great, but fascinating facts and footage of sequoias, waterfalls, and the Grand Canyon are. A fine way to vacation plan or at least armchair travel.

* **Kids Explore Kenya** To research a movie on this African country, the club visits a library to read about its history, an archaeologist's lab to learn about our oldest ancestors discovered there, and a theater to experience Kenyan music and dance. A school fair nicely presents today's culture in food, clothing, stories, and education. Without sugarcoating the country's problems of overpopulation, the video offers a fun way to learn about Kenya and its people.

* **Kids Explore Mexico** The club learns all about our Central American neighbors when they receive a package from their Mexican pen pal. As they pull out a sombrero, serape, mask, etc., a letter explains each item's place in the culture. Films of Mexico's regions and visits to a Mexican restaurant, museum, and festival round out the video.

***Who Will Be My Friend?** (Golden, 25 min., ages 3–6) Meet Molly and Ollie in this excellent antidote for first-day-of-school jitters. Molly is the little girl whose nervousness is assuaged by her toy dinosaur Ollie. He comes to life in several animated sequences to encourage her in song. When Ollie is missing from her cubby and she needs help in finding him, Molly asks the title question. Appropriately, the pace is unhurried, with enough story, song, and animation to keep kids' interest.

***World's Greatest Dinosaur Video, The** (MPI, 80 min., ages 4–10) A triple feature hosted by Gary Owens and Eric Boardman, who make learning fun by injecting a little humor. Despite its length, the tape moves with a magazine format loosely connected by storyline. It's like a video safari combining live-action, animation, and vintage film clips. Visits include a museum with moving models, the La Brea Tar Pits, and Dinosaur National Monument. And interviewed experts relay the latest information and speculation. Well-paced to cover both scientific fact and animated fancy of what the beasts looked like, *The World's Greatest Dinosaur Video* lives up to its name. See also its companion tape, *Son of Dinosaurs*.

FAMILY TOPICS

These are videos you should watch with your children, and you'll be happy to do so. Entertaining, and often covering important topics, they should spark good family discussions afterward. You'll find these videos organized into: Dramas and Documentaries, Drugs, the Environment, and Sex Education.

DRAMAS AND DOCUMENTARIES

African Journey (Public Media, 3 hrs., ages 8–14) WonderWorks title in which Canadian Luke joins his miner father in Africa. There he becomes friends with Themba and his family. Their clash and connection form the heart of the story, as each comes to question his own culture's ways. There's some great location footage including a safari, and a good subplot follows Themba's sister's desire for self-determination. A slow starter that's worth sticking with as the characters develop and relate.

Almost Partners (Public Media, 55 min., ages 8–14) Paul Sorvino stars in this WonderWorks mystery of Who Stole the Urn? Frustrated that he's not working on the big art theft, Sorvino discovers that he *is*, thanks to the persistence of his "assistant" Molly. How the $3 million Ming vase became her grandpa's urn is part of the mystery. With bits of humor and lots of deduction, *Partners* lays out and pieces together its clues effectively.

Amazing Stories, Book One (MCA, 70 min., ages 8–14) Two episodes of the Steven Spielberg series featuring Hollywood's best actors and production teams. *The Mission* stars Kevin Costner as a World War II flyer and Kiefer Sutherland as his young radio operator. The atmosphere is well set, as are the characters, filming, and suspense of knowing that something's going to happen, but what? And the story is a knockout, with the heroism and camaraderie of wartime and its—well, amazing ending. Danny De Vito and Rhea Perlman co-star in *The Wedding Ring,* a comedic and macabre tale of a ring's deadly effects.

*__And the Children Shall Lead__ (Public Media, 58 min., ages 7–12) WonderWorks title that dramatizes Mississippi's 1964 black voter registration in the wake of the Civil Rights Act. The players are young and old members of two families, one black and one white, with LeVar Burton as an organizing Freedom Rider. As the adults illustrate the diversity of attitudes toward change, their children struggle to understand what's going on. Danny Glover leads an excellent cast in this story that draws on period television footage for historical context.

Astronomers, The—Volumes 1–6 (PBS/Pacific Arts, 6 hrs., ages 8 and up) Five years and $5 million in the making, this 1991 PBS series looks at astronomy's most recent developments. Some episodes get bogged down in astronomer profiles, but two stand out as entertaining. *A Window to Creation* explores attempts to solve the mystery of how galaxies formed from the gaseous state following the Big Bang. And *Stardust* features wonderful animation of galaxies in motion and of stars forming and exploding.

*__Beethoven Lives Upstairs__ (The Children's Group, 52 min., ages 7–12) Another winner from the creators of **Raffi in Concert with the Rise and Shine Band.** The live-action, fictionalized account tells of ten-year-old Christoph's inital irritation at his home's new tenant, then fascination with the composer's eccentricity, moving to friendship with the man and admiration of the artist. It's an engaging story with a good sense of Beethoven's demanding genius and frustration at his growing deafness. Set in Old Vienna, *Upstairs* is richly appointed, filmed, and scored with Beethoven's own music. The premiere performance of the Ninth Symphony, attended by Christoph and his mother, proves a stirring climax.

*__Bridge to Terabithia__ (Public Media, 55 min., ages 10–14) Based on Katherine Paterson's Newbery Award-winning book, *Bridge* offers a realistic and moving tale of death. A lonely twelve-year-old boy and girl find each other and share an imaginary land called Terabithia. It's an engrossing story—first of their friendship, then of his recovery from her accidental death. Starring Annette O'Toole as his caring teacher, *Terabithia* is peopled with warm, likable characters. Their interplay saves this drama from being a tearjerker. It's more an ultimately uplifting tear*earner*.

*__Brother Future__ (Public Media, 110 min., ages 8–14) WonderWorks feature starring Phill Lewis as T. J., a hip-hoppin', street rappin' kid from Detroit who suddenly finds himself a slave in 1820s South Carolina. T. J.'s story has drama as a slave rebellion unfolds, humor as he slams Lifestyles of the Rich and Racist, and mystery as he tries to get back to the future. Naturally, it also has some violence, in a couple slave-whipping scenes. Lewis is an engaging actor as he displays a range of emotions in trying to apply today's ideas to yesterday's realities. A gripping story.

*__Civil War, The—Volumes 1–9__ (PBS/Pacific Arts, 11 hrs., ages 11 and up) A media event when first aired in 1990, Ken Burns's series paints a detailed, horrifying, and very human portrait of America's most destructive war. Set to poignant period music and filled with compelling storytelling, *Civil War* remembers the story in history. And yet it's herstory, too, as one of the most eloquent voices is that of Mary Chesnut. Other voices heard include Walt Whitman, President Lincoln, generals from North and South, with soldiers themselves the most affecting. Their letters and journals, along with the stories of major battles and strategies, form the heart of the program.

With a well-paced variety of visuals—photographs, prints, location shots, maps, and historian interviews—its length is not a problem, especially if you pace your viewing. A good episode to start with is Volume 3, *Forever Free,* set in the summer of 1862. It tells the story of Antietam, a triple battle with the dubious distinction of claiming the bloodiest day in American history. And it does a marvelous job of distilling a multitude of facts and experiences into an insightful and coherent whole. If you think your child is ready for it, *Civil War* is a fascinating history lesson.

Dirkham Detective Agency, The (Lorimar, 45 min., ages 4–7) Excellent beginner mystery from Scholastic. The kids who run the agency are out to prove they're good detectives in their first case—a dognapping. Starring smart yet realistic kids, the video includes other minimysteries to solve. And the atmosphere is nonthreatening, making it good fare for the littlest sleuths. See also *Mystery at Fire Island.*

Divorce Can Happen to the Nicest People (LCS/R&G, 30 min., ages 5–10) Animated self-help video from the creators of *Where Did I Come From?* and *What's Happening to Me?* Like those, *Divorce* explores the situation with candor, light humor, and reassurance that you're not alone. Despite some sexism—women don't like rock climbing nor men aerobics, and its implication of there being men's vs. women's work—the video offers a well-reasoned and illustrated account of why parents divorce. Kids learn it's not any one person's fault and that they are still loved by both parents.

Frog Girl: The Jenifer Graham Story (David Eagle, 47 min., ages 8–14) CBS Schoolbreak Special based on a true story. When Jenifer Graham, a tenth grader, asks to be excused from dissecting a frog, the principal tells her to get a lawyer. She does, and her cause against animal cruelty is well made. Jenifer gives a particularly eloquent defense of her position at a state hearing. An absorbing and honest drama that presents both sides fairly.

Growing Up in the Age of AIDS (MPI, 70 min., ages 11 and up) *When* you watch this video with your child will depend on his or her readiness for a frank discussion of sex. But watch it you should, as it presents to teenagers the facts about AIDS. The ABC News Special is hosted by Peter Jennings, whose on-stage audience has parents, teachers, doctors, including Surgeon General Antonia Novello, and mainly teens themselves, some of whom are HIV-positive. The emphasis is on prevention and protection, with experts giving the latest research findings to answer teens' questions. The dialogue begun on this tape should segue well to one within your family.

Gryphon (Public Media, 55 min., ages 6–10) As a substitute teacher teaching "substitute" facts, Amanda Plummer mesmerizes in this WonderWorks video. Her magical storytelling, including that of the mythical gryphon, inspires her inner-city students to better their lives. Their creation of a mural and learning to judge people, not on appearances but on what they do, show how good education can be with the right spirit.

***Jacob Have I Loved** (Public Media, 55 min., ages 10–14) WonderWorks video based on the Newbery Award-winning novel by Katherine Paterson. Bridget Fonda stars as the sixteen-year-old Louise, whose beautiful and talented twin sister gets all the attention, not only from her family but everyone on the island. Everyone except Captain Wallace, that is, who likens his friend Louise to a crab, with "a hard shell and ready to bite." Relating his own experience, he helps her see that hatred and jealousy, however well-founded, hurt herself the most. And he encourages her to find the one gift that's hers. Fonda turns in a wonderful performance as the brittle but hopeful Louise. An all-around quality production guaranteed to move.

Little Princess, A—Volumes 1–3 (Public Media, 55 min. each, ages 6–10) Frances Hodgson Burnett's tale of Sara Crewe is brought to beautiful life in this WonderWorks video. As the stalwart young romantic, Amelia Shankley brings the right air to the role. Her Sara is genuinely charming whether up or down in her fortunes. As the once and future "princess," Sara maintains her faith in herself and others by imagining she's a good-hearted princess. Sara's is an engaging story, full of Victorian sentiment and irony and yet never sinking to melodrama. With period costumes and customs, *Little Princess* is a joy to watch.

Maricela (Public Media, 55 min., ages 6–10) A WonderWorks story of a thirteen-year-old El Salvador émigré trying to adjust to life in the U.S. Maricela and her mom live with Linda Lavin and her teenage daughter. When the girls' relationship strains, Maricela takes off but is followed by the curious Stacy. An original story whose themes touch on prejudice, feminism, and appreciation of cultural diversity.

Mighty Pawns, The (Public Media, 55 min., ages 7–14) Based on a true story, this WonderWorks video stars Alphonso Ribeiro and Paul Winfield in the tale of a junior-high chess team. Ribeiro is one of a group of toughs who form the team after losing a bet to their detention teacher, played by Terence Knox. Theirs is an absorbing drama, dense with points about using your intelligence, giving people a chance, and competing as better than fighting. These values are artfully woven into the story, which also benefits from fine performances.

Mystery at Fire Island (Lorimar, 45 min., ages 7–12) Two young cousins investigate the disappearance of an old fisherman. Dash is an artist whose broken leg keeps her in a wheelchair, and Jess a collector of sound effects on his tape recorder. Together they figure things out and with enough time between clues and solutions that the viewer can, too. Fine Scholastic production that's loaded with atmosphere. See also *The Dirkham Detective Agency*.

Red Balloon, The (Nelson, 34 min., ages 3–10) Awarded by both the Academy of Motion Picture Arts and Sciences and the American Film Institute, Albert Lamorisse's 1956 film is timeless. The balloon in question resolutely follows a little boy around Paris after he releases it from its "trap." It becomes an insistently colorful icon in every scene, symbolizing children's innocence, joy, obviousness, and need for attention. Is *Red Balloon* a paean to childhood? Paris? Silent movies? Take your pick, it works on all levels and for all ages.

Roots, Volumes 1–6 (Warner, 2 hrs. each, ages 10 and up) Credited with sparking a genealogical revolution, this 1977 miniseries remains the highest-rated of all time and possibly the most-awarded with over 145, including nine Emmys. Alex Haley's search for his "Roots" yielded a family history to be shared by countless African-Americans denied their heritage. Opening in West Africa in 1750, the first story is that of Kunta Kinte, played by LeVar Burton. His giving up his only chance of escape for the sake of his daughter sets the stage for the 100 years drama. Each generation finds its own tensions between family and self, family and a "higher" good, and doing what's right vs. what's smart. Ably played by a cast including Cicely Tyson, Ed Asner, and Ben Vereen, *Roots* tells this American his-and-herstory with honest and at times intense emotion.

Runaway (Public Media, 55 min., ages 10–14) Jasmine Guy stars in this WonderWorks video about a thirteen-year-old boy on his own in New York. Living in a subway, Aremis makes friends with Guy, a waitress, and a Vietnam vet artist. Each has a problem to resolve, which is addressed through their growing affection for each other. *Runaway* is a poignant drama that shows the realities of city life—its homeless, gangs, children, *and* good-hearted people.

Speeches of Martin Luther King, Jr., The (MPI, 60 min., ages 7 and up) Collection of six speeches that charged the civil rights movement. Presented just as they were given, they reveal Dr. King's strength, warmth, humor, and persuasiveness. Most powerful are his defense of nonviolence as an effective method of change and the famed "I have a dream" speech.

*__Sweet 15__ (Public Media, 2 hrs. ages 10–14) WonderWorks drama of Marta's Quinceañera, the Hispanic celebration of a girl's 15th birthday, and its unsettling by her father's being an illegal alien. Karla Montana shines as Marta, a Mexican-American girl who is forced to see that her father's problems are more important that her coming-of-age party. Helping him, in fact, shows her coming of age more significantly than any party could.

On the cusp of childhood and adulthood, Mexican and American cultures, Marta has a story as entertaining as it is illuminating. A beautiful and touching film.

*__Tommy Tricker and the Stamp Traveller__ (FHE, 101 min., ages 6–10) Title character *is* a great trickster whose swindle of a young stamp collector sends both on an adventure around the globe. They manage this by shrinking down into stamps and letting the postal service transport them. Their "vehicles" include a Canadian Mountie's horse, a Chinese dragon kite, and an Australian kangaroo—all on stamps. A fine adventure well told for the whole family.

When Mom and Dad Break Up (Paramount, 30 min., ages 6–14) Host Alan Thicke opens by talking with kids of divorcing parents about their confusion, disbelief, and worry. Then songs, animation, and advice from Thicke explore the feelings a family encounters as it dissolves and adjusts. The effect of the tape parallels that of reality. . . a process from pain to recovery. Except for the fact it assumes dad is angry and leaving, *Break Up* is an excellent tape that deals with an emotional issue in a straightforward, and yet warm manner.

Woman's Place, A (View, 25 min., ages 8–14) Of course, it's not necessarily in the kitchen. Julie Harris narrates this Time documentary of women in the arts, science, business, and sports. Included are such luminaries as Helen Keller, Barbara Jordan, Susan B. Anthony, Margaret Mead, Joan Ganz Cooney, Harriet Tubman, and Wilma Rudolph. While their bios are a little too rapid-fire, the video provides a good overview of women's achievements in recent history.

You Must Remember This (Public Media, 110 min., ages 7–12) Robert Guillaume stars in this WonderWorks feature about "the first Spike Lee." Sharing a love of old movies with his niece Ella, Guillaume suddenly withdraws after a trunk arrives from an old friend. Ella discovers his past as an early black filmmaker through conversations with a movie projectionist and a researcher with the American Film Institute. While the program could have been shorter, *Remember* is entertaining for Guillaume's performance, Ella's solving the mystery of the trunk, and an illuminating segment on "Hollywood's Hall of Shame" regarding its characterizations of African-Americans.

DRUGS

Cartoon All-Stars to the Rescue (Walt Disney, 30 min., ages 6–10) Recognizing the importance of saying "no" at an early age, artists and corporations, through the Academy of Television Arts and Sciences, created this antidrug special. Its all-star lineup includes Bugs Bunny, Teenage Mutant Ninja Turtles, Winnie-the-Pooh, Garfield, and the Muppet Babies. All the right messages come through, woven into the story of a drug abuser and his younger sister. Particularly artful are metaphors for drug abuse, such as crazy mirrors for its distortion.

Drug Free Kids: A Parents' Guide (Fusion, 70 min., ages 8–14) Best all-around video on the topic is hosted by Ken Howard and sponsored by the Scott Newman Foundation. Where others may use scare tactics and push rehab centers, *Drug Free* encourages parents to take responsibility for their children's lives. Credible and entertaining vignettes show how to get back in touch and in charge. Preventive advice is also excellent, making this tape worthwhile to parents of young children as well.

Nightmare on Drug Street (35 min., ages 10–14) At press time, this award-winning video was out of release, but you may still find it on some shelves. Avoiding the shock approach, it nevertheless conveys drugs' dangers in three dramatic miniplays. Felipe, a high school senior, drinks and smokes his way to a fatal car crash. Fourteen-year-old Jill only compounds her lonely feelings with cocaine. And Eddie, a twelve-year-old with a hidden heart problem, pays the ultimate price for trying crack. The stories are well done to show, not just tell, how drugs can change you for the worse.

THE ENVIRONMENT

Berenstain Bears and the Trouble with Friends, The (Random House, 30 min., ages 3–7) In the title story, Sister Bear finds her new friend fun but "a little bossy and a little braggy." After a talk with Mama, Sister realizes we all have to tolerate friends' weaknesses. In the video's second story, *The Coughing Catfish,* the Bear family learns why Jake can't breathe. A hot-air balloon ride reveals the source of his lake's pollution, and an underwater voyage its solution.

*__Dr. Seuss: The Lorax__ (CBS-Fox, 30 min., ages 4–10) Ted Geisel, master at distilling situations to their essential truths, turns to the environment in this video. As the forest's "lone voice for the voiceless," the Lorax stars in this fable of ecological disaster. The Once-ler, driven by ignorance and greed, fails to heed the Lorax and succeeds in eradicating the Truffula Tree. Powerfully imaged, with vast scenes of tree stumps and garbage, *Lorax* ends hopefully, rather than happily.

*__Eco, You, and Simon, Too__ (3E Communications, 40 min., ages 2–6) A great first tape about the environment, with a colorful collection of songs, rhymes, and games. Eco is a sea otter puppet, and Simon his childlike adult friend. Together they engage "you," the young viewer, in education about food, flowers, animals, and trees. Bits of imaginative animation and movement keep *Eco* highly entertaining.

FernGully... The Last Rainforest (CBS-Fox, 70 min., ages 3–8) Based on stories by Australian Diana Young, this animated feature crytallizes environmental issues for young children. Marking trees for destruction, Zak is shrunk by a fairy to save him from a falling tree. Crysta then shares the wonders of her world, recruiting him to help save FernGully. Like the rainforest itself, the movie has rich tones, varied settings, and a colorful cast, including a wacky bat voiced by Robin Williams. And the contrast struck between the beautiful and natural magic of the fairies vs. the mindless and mechanical rapaciousness of developers is forceful without frightening young children. Despite lackluster animation and music, *FernGully* makes a sweet and coherent intro to eco consciousness.

50 Simple Things Kids Can Do to Save the Earth (David Eagle, 47 min., ages 6–14) CBS Schoolbreak Special hosted by Brian Austin Green and Sara Gilbert. Like others, this video is jam-packed with ideas your family can do at home—recycling, planting trees, and turning off lights. What's great about this video, though, is that it shows groups of kids in environmental action. One group patrols a beach for garbage, another is cleaning up a creek. The Philadelphia Garden Project is reclaiming city parks from drug dealers by planting and caring for vegetables, flowers, and trees. Yet another group visits Costa Rica each year to help sea turtles and their eggs. An inspiring video to raise eco consciousness and action.

Gift of the Whales (Miramar, 30 min., ages 7–12) Educational drama about a Native American boy whose life is turned around by a visiting scientist. Learning about whales from him and his grandfather, Dan determines to become a whale scientist, too. The teacher's enthusiasm and affection for this "greatest gift of the sea" passes on to the viewer, too. With endangered statistics and beautiful whale footage, *Gift* imparts a sense of us all sharing the Earth. See also *Spirit of the Eagle.*

Help Save Planet Earth (MCA, 71 min., ages 6–14) Hosted by Ted Danson, this video stars Jamie Lee Curtis, Whoopi Goldberg, Sinbad, and Beau and Lloyd Bridges. Entertaining vignettes hone environmental themes on toxics, recycling, water conservation, energy efficiency, and endangered species. Danson introduces each skit and closes with suggestions for getting involved. A great family video full of workable ideas.

It Zwibble: Earth Day Birthday (FHE, 30 min., ages 3–8) Directed by Michael Sporn, *Zwibble* is drawn and animated with characteristic charm. Zwibbles are young dinosaurs who've magically come to life only recently, and It Zwibble is their "fairy godfather." Young Orbit, after a trip around the world showing its beauty and destruction, suggests giving Earth a birthday party. Beautifully colored and full of sparkly touches, the animation's point is that in caring for the Earth, no task is too great or too small.

Pete Seeger's Family Concert —see Music

Samson and Sally: The Song of the Whales (Celebrity, 70 min., ages 4–9) Animation of a Ben Haller book that runs the gamut from drama to humor and beauty to horror. It chronicles the adventures of two young whales as they search for Moby Dick. Because many of their "adventures" result from human abuse—harpooning, oil slicks, and nuclear-waste dumping—the video makes a good case for the environment. There are some playful moments, but the dangers' realism behooves you to watch with your child.

Spirit of the Eagle (Miramar, 30 min., ages 5–10) From the creators of *Gift of the Whales* comes another nature/coming of age video. On a field trip to the mountains, a boy and girl learn about eagles in a way that encourages their own development. The *Spirit of the Eagle* is to try hard and survive despite the obstacles. For the bald eagle these obstacles include pesticides and land development, making this a good environmental tape. With amazing footage of this largest raptor bird, *Spirit* soars and well conveys a consciousness of animals living in the wild.

SuperTed: Leave It to Space Beavers (Hanna-Barbera, 90 min., ages 5–9) SuperTed is a wonderfully British teddy bear/super hero who conquers evil with more brains than brawn. This animated video of four stories includes the title episode of environmental theme. Dr. Frost has imported alien beavers to destroy the world's forests. SuperTed and sidekick Spotty draw the beavers to their side and "cream" Frost with tutti-frutti ice cream. Shot through with humor, SuperTed is well drawn and succeeds with nonviolent cleverness.

3–2–1 Contact videos (Children's Television Workshop, ages 6–14) The award-winning CTW science program has four environmental "extras" on video. Each offers an entertaining blend of location footage, animation, music, demonstrations, and charts to educate on an ecological issue. And host Stephanie Yu does a fine job of pulling it all together. Watch with the kids for an education of your own.

Bottom of the Barrel (30 min.) Our dependence on oil for transportation, heating, and many plastics is shown along with its dangers of air pollution, oil spills, and shrinking supplies. Best segments include a mock spill and kids' attempts at clean-up, time-travel animation to show oil's creation, and energy alternatives.

* **Down the Drain** (30 min.) It's amazing how much education CTW is able to convey in a half-hour. This special focuses on water and our problems with its quantity and quality. You'll also learn about water's unique properties, the water cycle in animation, how treatment fits into that cycle, and how to conserve this precious resource.

* **The Rotten Truth** (30 min.) The Rotten Truth about garbage is that it doesn't go away just because we throw it away. To sort out a complicated problem, CTW visits a landfill, demonstrates that you can't make nothing out of something, and shows that recycling is only natural through time-lapse photography of decaying fruit.

* **You Can't Grow Home Again** (60 min.) Our endangered rainforests are featured in this video, which includes a trip to Costa Rica. So loaded with life are these special places that many of their species have yet to be named. Concepts, such as biodiversity and species extinction, are well illustrated. And several miniquizzes point up the magnitude and urgency of the problem.

Time Warner Presents the Earth Day Special (Warner, 95 min., ages 8–14) Mother Earth, played by Bette Midler, is critically ill in this special from April 1990, the 20th anniversary of the first Earth Day. Her diagnosis, suggested treatment, and prognosis are delivered by the likes of Robin Williams, Candice Bergen, Carl Sagan, Dustin Hoffman, D.J. Jazzy Jeff, and the Golden Girls. Leaning more toward entertainment than education, *Earth Day* nonetheless makes a good intro to our environmental problems.

***Widget's Great Whale Adventure** (FHE, 47 min., ages 6–10) Animation starring an eco-conscious, shape-shifting alien. In this adventure, Widget helps teen Earthers save whales from illegal hunters. In the second story, *Gorilla My Dreams,* he learns about zoos and pets. Kids loves the silly mistakes Widget makes, even as they learn about our environmental errors. Also available is *Widget of the Jungle.*

***World Alive, A** (Sea Studios, 25 min., ages 3–14) All creatures great and small star in this stunning paean to Earth and its inhabitants. Narration by James Earl Jones often gives way to music and ambient sound to tell the story. And the story is glorious, showing nature in all its colors and moods. Children will be most entranced by the variety of wildlife shown at work and play. Its amazing footage, on land and sea, well portrays Earth as a "fragile living tapestry that only we can unravel or preserve."

Yakety-Yak—Take It Back! (Take It Back Foundation, 50 min., ages 8–14) This is the video *behind* the music video pulling together nineteen artists to promote recycling. The likes of Natalie Cole, Quincy Jones, Bette Midler, and Stevie Wonder chime in with Bugs Bunny and Skat Cat in a parody of the 50s classic. In addition to the clip are twenty-five public service announcements the celebrities made and a making-of-the-video segment. A hip-hop video to get you and yours to reduce, re-use, and recycle.

SEX EDUCATION

***Miracle of Life, The** (Crown, 60 min., ages 10 and up) This Emmy Award-winning "Nova" program is a miracle itself. In 1982, Swedish filmmakers recorded the act of conception on film. The result is a flowing work of art worthy of a miracle's presentation. From the release of the egg and the sperms' amazing journey, the story of human conception continues through birth. An excellent beginning in sex education.

***3–2–1 Contact: What Kids Want to Know About Sex and Growing Up** (Pacific Arts, 60 min., ages 10–15) With characteristic skill, *3–2–1 Contact* covers the areas of puberty and sex in this special now on video. At its heart are group sessions with sex educators Rhonda Wise and Robert Selverstone. They earn an A+ for their explanations, which are straightforward, complete, and reassuring. Opening with a segment on Boys and Hormones, the tape moves on to Girls and Hormones, talks about Puberty and Parents, and then Sex and Its Responsibilities. Well worth watching yourself first and then sharing with your child when he or she is ready.

What's Happening to Me? (R&G, 30 min., ages 10–15) Like the Peter Mayle and Arthur Robins book, this animated video explores the changes of puberty with candor and humor. Information-rich, its presentation is logical, clever, and reassuring. All the whats, whys, and wherefores are covered on such topics as hormones, genital development, acne, and menstruation.

Where Did I Come From? (R&G, 30 min., ages 6–12) Animating the best-selling book by Peter Mayle and Arthur Robins, this video is well-written, accurate, and funny. Its claim that the truth is far more interesting than reproduction myths finds support in clear instruction and amusing animation. The sex act itself is defined with remarkable honesty and yet on a level suitable even for young children.

FOLK AND FAIRY TALES

These videos are personal favorites. They draw not only on the rich European tradition of fairy tales, but also on the larger, richer legacy of childhood stories from around the world. Each has a special message, unique characters, and a story that reflects its origins and yet sparks a universal response. Enjoy!

Aesop's Fables: The Hen With the Golden Eggs (Vestron/FHE, 50 min., ages 6–10) Italian animation of nine fables, including *The Lion in Love, The Dog and His Image,* and *The Fox and the Crow.* They're filled with good tricks and poetic justice for the foolish, the greedy, and the vain. Rewritten for contemporary ears and painted in Rouault-like colors, the fables are also nicely animated. So they're as entertaining today as the long-ago day they were first spoken.

***Anansi** (Rabbit Ears, 30 min., ages 3–10) From We All Have Tales, a Jamaican legend of a clever spider. The best of two stories tells of Anansi's tricking a snake into captivity. Denzel Washington makes a wonderful storyteller, adopting all the voices of the jungle. Music from reggae band UB40 lends the right island sound. Best of all is Steven Guarnaccia's art of brilliant aquas, reds, and greens on brown paper. While only slightly animated, *Anansi* moves with these elements and entertains a wide age range of kids . . . and adults.

Annie Oakley (CBS-Fox, 53 min., ages 6–12) Shelley Duvall's Tall Tales and Legends entry starring Jamie Lee Curtis as "the truest of all American legends." A woman of determination who could always hold her own, Oakley's story dramatizes some common hurdles. Not only is she discouraged from unladylike activities, she overcomes the resistance to admit women into traditionally male areas. An entertaining look at an early showbiz star who meets the likes of Sitting Bull, Mark Twain, and Queen Victoria.

Annie Oakley (Rabbit Ears, 30 min., ages 6–12) From the American Heroes & Legends series, Annie's story is told by Keith Carradine, reprising his Broadway role as Will Rogers. The video opens with Oakley's stardom in Buffalo Bill Cody's Wild West show and then flashes back to her childhood and climb to fame. Admired by everyone from European royalty to the Sioux people, who dubbed her Little Sure Shot, Oakley was possibly the best-known woman of her time. This video shows why, with watercolors from Fred Warter and music from Los Lobos that hit the mark as surely as their sharpshooting subject.

Beauty and the Beast (CBS-Fox, 50 min., ages 6–10) For a live-action *Beauty,* try this Faerie Tale Theatre production starring Susan Sarandon, Klaus Kinski, and Angelica Huston as one of the envious sisters. With exquisite costumes, sets, and music, the video holds magic at every turn. And the storytelling is spellbinding, despite its sappy ending. A classic retelling that manages suspense in a known tale.

*__Beauty and the Beast__ (Lightyear Entertainment, 30 min., ages 5–12) As the premiere release of Stories to Remember, the romatic fantasy is treated to stunning artwork by Mordicai Gerstein and score to match by Ernest Troost. And Mia Farrow makes the perfect reader for its poetic language. While only slightly animated, the video's filled with magical moments as Beauty comes to trust her heart and not her eyes. A delight for the whole family.

*__Beauty and the Beast__ (Walt Disney, 80 min., ages 3–10) An instant—and flawless—classic. The first animation to garner an Academy Award nomination for best picture, *Beauty and the Beast* boasts wonderful art, animation, characters, Oscar-winning music, storytelling, and humor. As if that weren't enough, it also encourages pro-intellectual, antisexist values of seeing past the surface. And how refreshing to have a brown-eyed brunette heroine! From the opening scene of the magical, floating rose, to Belle's rousing introduction, to Beauty and the Beast's starlit ballroom dance, to the slapstick finale battle, the film entrances again and again. A family treat to be enjoyed again and again.

*__Boy Who Drew Cats, The__ (Rabbit Ears, 30 min., ages 6–10) We All Have Tales video that draws on many talents. William Hurt gives a mesmerizing reading of the Japanese folktale. Drawings by David Johnson are no less than fine art, delicately-toned with Oriental lines. And music from Grammy-winning Mark Isham tells a story of its own. The boy's story is one of a drawing obsession that proves useful against a terrible demon. Because of its menacing atmosphere and somewhat bloody climax, this video is best reserved for older children.

*__Br'er Rabbit and the Wonderful Tar Baby__ (SVS, 30 min., ages 3–10) Rabbit Ears Storybook Classic read by Danny Glover, illustrated by Henrik Drescher, with music by Taj Mahal. The American folktale recounts Br'er Fox's attempts to "fricassee" Br'er Rabbit. But he's too clever for his own good in this well-done dissolve animation. Drescher's artwork of luminous, surprising colors echos American naïve paintings. And Glover makes a great storyteller, relishing its twists and turns of phrase. A perfect video for Br'er Rabbit and his "well-known sass."

*__Cinderella__ (Walt Disney, 78 min., ages 3–10) Disney magic at its best. The animation is superb, from the flow of ribbon in the dressmaking scene to Cinderella's glamorous transformation. The story has comedy, heart, and drama, with an ending as emotionally satisfying as any. Add the antics of Cinderella's mice friends and a memorable score, and you have a near-perfect film.

Cinderella (CBS-Fox, 84 min., ages 5–10) Rodgers and Hammerstein musical that aired on television in 1964. In her debut role, Lesley Ann Warren makes a sweet and radiant Cinderella. Celeste Holm plays her fairy godmother, Ginger Rogers and

Walter Pidgeon the queen and king, and the forgettable Stuart Damon her prince. While some numbers are fast-forwardable, the songs "Impossible!," "Ten Minutes Ago," and "Are You Beautiful?" add musical magic to the fairy tale spell.

Cindy Eller (Strand, 44 min., ages 5–10) An ABC Kidtime Special, this video updates Cinderella believably and entertainingly. Cindy, played by Kyra Sedgwick, has just moved in with her father, his new wife played by Melanie Mayron, and two stepsisters, including Jennifer Grey who nearly steals the show. Pearl Bailey plays Cindy's fairy shopping-cart lady with finesse. And the story manages to stay true to the original while holding a few surprises.

Dancing Princesses, The (CBS-Fox, 50 min., ages 6–10) Lesley Ann Warren shines as Jeanetta in this Faerie Tale Theatre adaptation of a Brothers Grimm story. With her sisters, she confounds her father with worn-out shoes each morning. Only Peter Weller, aided by a magic cloak of invisibility, can discover the reason. Filled with festivity in moonlit dances and feminism in Jeanetta's refusal to be "some prince's door prize," *Princesses* is a royal treat.

Darlin' Clementine (CBS-Fox, 50 min., ages 6–10) Shelley Duvall's Tall Tales and Legends video starring its producer. The story behind the song takes place in the Great Gold Rush, when men left their homes and families in search of fortune. The only woman in a mining camp, Clementine tries to convince the miners that dreams are more important than gold. Infused with humor, special effects, and sound storytelling, Clementine's tale is one of good-heartedness in the face of greed. Randy Newman sings title song at the end.

Elves and the Shoemaker, The (Hanna-Barbera, 30 min., ages 3–8) A Hallmark Timeless Tale animating a Brothers Grimm classic. The shoemaker and his wife make shoes that last so long their business is off. Then, in step some magical elves to fashion shoes so fancy they'll bring in new business. And they do, until a bothersome cat chases the elves away. With the help of a Scooby-like dog, the elves prevail, of course, to a happy ending. The characters are cute and well animated, especially in the magical shoemaking scenes.

*__Emperor and the Nightingale, The__ (SVS, 40 min., ages 4–10) Rabbit Ears Storybook Classic which draws on excellence from many quarters. Glenn Close beautifully reads the Hans Christian Andersen tale of the wonders of nature and love. Grammy composer Mark Isham contributes an Asian-influenced New Age score. And the drawings of Robert Van Nutt are exquisite—frame after frame of Oriental art. This *Nightingale* is indeed a treasure.

*__Emperor's New Clothes, The__ (SVS, 30 min., ages 3–10) John Gielgud perfectly reads this Rabbit Ears Storybook Classic of the Hans Christian Andersen tale. Like Goya's cartoons, Robert Van Nutt's artwork has a classic look with fresh colors and lighting. Mark Isham turns in a lush Baroque score to reflect the pomp and delicacy of the times. And adaptation by Eric Metaxas emphasizes the humor of a hoax built on shared foolishness. A funny family delight told in dissolve animation.

*__Emperor's New Clothes and Other Folk Tales, The__ (Children's Circle, 30 min., ages 3–10) This Gene Deitch animation draws extra smiles from an already hilarious tale. The errant tailors resemble flim-flam men, the Emperor fancies a Superman outfit, and Dixieland jazz keeps things moving. Bernard Westcott's original art, bright and colorfully patterned, is a standout. *Why Mosquitoes Buzz in People's Ears*, also animated by Deitch, tells the African folktale in poetic language and gorgeous cloisonné-like art by Leo and Diane Dillon. Wrapping up the video is *Suho and the White Horse*, a suspenseful Mongolian tale of everlasting friendship.

Faerie Tale Theatre Shelley Duvall, a collector of fairy-tale books, produced this series of live-action videos. Recruiting her talented friends to the project, she was able to recreate the classics for a new generation. Well acted, staged, and written with bits of contemporary humor, Faerie Tale Theatre is family entertainment at its best. Reviewed in this section are *Beauty and the Beast, The Dancing Princesses, Little Red Riding Hood, Rapunzel, Snow White and the Seven Dwarfs, The Tale of the Frog Prince,* and in the Holidays: Halloween section, *The Boy Who Left Home to Find Out About the Shivers.*

Finn McCoul (Rabbit Ears, 30 min., ages 4–10) Catherine O'Hara tells this We All Have Tales story in fine Irish brogue. Equally wonderful is music from Boys of the Lough, adding its own lilt to the legend as told in dissolve animation. It would seem a clash of the titans is coming, as the monstrous Cucullin seeks out Finn, also a giant. But Finn's wife Oonagh has other plans, and together they trick Cucullin in a way that's as funny as it is clever. An Irish tall tale well told in the expressive language of the land.

Fool and the Flying Ship, The (Rabbit Ears, 30 min., ages 6–10) Robin Williams narrates this Russian entry in the We All Have Tales line. When the czar decrees whoever can build a flying ship will win the hand of his daughter, the peasant fool is lucky enough to find one. On the way to claim his reward, the fool picks up friends who will later help him fulfill new, "impossible" requirements set by the czar. Figuring out who will perform which task is half the fun of the video. Henrik Drescher's art, with characters that have Mr. Potato Head-like parts, is bizarre but funny to kids. And Williams is his usual droll self, delivering lines like "penguin" being Eskimo for tuxedo.

Foxy Fables (Media, 75 min., ages 4–9) Fables from the likes of Aesop and Uncle Remus are treated to amusing clay animation. Fox the fox, Brixton the rabbit, Cranium the crow, et al. act out over a dozen stories. They tell of clever cons and just deserts, tug-of-wars and tar dolls. Since each lasts about five minutes, one or two would make good bedtime fare.

Jack and the Beanstalk (Hanna-Barbera, 51 min., ages 3–8) Emmy Award-winning adaptation directed by and starring Gene Kelly as a peddler friend of Jack's. With Hanna-Barbera's help, animation combines with live-action to heighten the tale's magical effects. Kelly dances through a starry sky and hides in the cracks of the giant's floor. While Jack himself is dated and corny, the tricks he and Kelly play on the animated giant are delightful.

Johnny Appleseed (CBS-Fox, 50 min. ages 5–10) One of Shelley Duvall's Tall Tales and Legends starring Martin Short, Rob Reiner, and Molly Ringwald. Against his parents' wishes, Johnny, played by Short, strikes out on his own—traveling, planting, and teaching people about growing things. Since he can talk to animals, Johnny enlists their help in his planting plans. When the beaver agrees to build a dam to reroute the river, Short does a patented Ed Grimley dance of joy. While it runs a bit long, *Appleseed* well reflects traditonal American values of enterprise and individualism and the renewed value of planting trees.

King Midas and the Golden Touch (Rabbit Ears, 30 min., age 4–10) Representing Greece in the We All Have Tales line, *Midas* is read with relish by Michael Caine. The story, brilliantly adapted by Eric Metaxas, traces the king's fatal attraction to gold. When the Midas touch turns even his daughter to gold, his real tears bring redemption. This magical tale uses dissolve animation of art evoking Greek classicism and features an original score from Ellis Marsalis and Yo-Yo Ma.

***Koi and the Kola Nuts** (Rabbit Ears, 30 min., ages 4–10) From We All Have Tales, Whoopi Goldberg narrates this African story with gusto. Denied his birthright, Koi sets off to find a village where he'll be treated like a chief's son. His generosity with his only possessions, the kola nuts, is rewarded as he tries to pass impossible tests in his new village. Another gem from Rabbit Ears, *Koi* is well written with suspense and humor, set to dynamic Herbie Hancock music, and strongly imaged in neon-bright tropical colors and patterns.

Little Mermaid, The (Starmaker, 75 min., ages 6–10) Look to this video for an adaptation of the Hans Christian Andersen story that's more faithful than Disney's. Without song interludes and side characters, the love story is more developed. And of course, it does not end happily ever after. This slightly animated *Mer-*

maid is nicely drawn, well written, and enhanced by lovely scenes of the night sea and ocean sunrises.

***Little Mermaid, The** (Walt Disney, 83 min., ages 3–9) From the moment Ariel swims into the scene, she hooks you into her story of longing to be human. As the daughter of the MerKing, however, she faces obstacles from him as well as his adversary, the Sea Witch. In Disney fashion, Ariel's adventures roll from the dramatic to the comic to the romantic, all with the promise of a happy ending. The animation, too, is Disney-best, whether in the realism of a sea storm or the loveliness of a moonlit serenade. A bubbly bauble of a video.

Little Red Riding Hood (CBS-Fox, 60 min., ages 3–8) Faerie Tale Theatre production starring a luminous Mary Steenburgen in the title role and a deliciously wicked Malcolm McDowell as the wolf. It's a comedic take on the classic, where the wolf reads "Eat Well and Feel Better" and Mary is an overprotected teen longing for adventure—or at least to be a cabinetmaker. Her suitor and savior, in fact, is an apprentice in that craft. A fun retelling that breathes new life into the old tale.

***Mickey and the Beanstalk** (Walt Disney, 29 min., ages 3–9) A Mini-Classic of maxi-entertainment. When Mickey Mouse brings home magic beans to a starving Goofy and Donald, the latter's reaction is the expected Duck á la rage. Mickey's vindication comes, of course, in the dramatic beanstalk growing which leads to their adventures in the giant's castle. In classic Disney fashion, the little guys cope comically yet successfully in their outsized surroundings. A singing Golden Harp enhances their chances as well as the cartoon's fairy-tale atmosphere. A family treat.

Mother Goose Stories (Lorimar, 30 min., ages 2–6) Jim Henson Play-Along video of three stories to watch and then act out with your own props. A Muppet Mother Goose introduces each by reciting the rhyme, and then a child and Muppet cast tells the story behind the rhyme. Little Miss Muffet was really just a poor little rich girl who wanted some fresh air. *Song of Sixpence* stars a poor little rich boy whose miserly aunt and uncle send out to make his own fortune. And Little Boy Blue is only sleeping because he was up all night with his sick mother. These simple renditions are well paced for little ones and offer something new to the classics.

*__Paul Bunyan__ (SVS, 30 min., ages 5–10) Rabbit Ears Storybook Classic that adapts the American tall tale literately and with a heap of humor. Jonathan Winters narrates in best old-timer voice. Leo Kottke turns in a pretty country score. And artwork by Rick Meyerowitz harkens back to early comics, with delights for the eye and funny bone. Conscientiously, the great logger ends his story in remorse and sets about planting as many trees as he felled.

*__Peachboy__ (Rabbit Ears, 30 min., ages 6–10) Sigourney Weaver turns in a spellbinding performance as narrator of this Japanese story in the We All Have Tales line. Marvelously born of his namesake fruit, Peachboy sets off to rescue children lost to the ogres. With beautiful watercolors and music from the composer of *The Last Emperor,* his quest becomes a moving work of art.

*__Pecos Bill__ (SVS, 30 min., ages 6–12) Rabbit Ears Storybook Classic pefectly narrated by Robin Williams. With art and humor as bright as the desert sun, this video paints a portrait of the man who put the "wild" in Wild West. Raised by coyotes, Bill just didn't know any better than to ride a cougar or grab a rattlesnake for his lariat. Full of rip-roarin' fun, it's a tall tale well told.

Peter Pan (GoodTimes, 104 min., ages 4–9) When this Broadway musical aired live in 1955, it was a television event. Mary Martin captured imaginations that night, as well as an Emmy for her wonderful performance. When color came to TV, *Peter Pan* returned in 1960. This video records the latter, color program in all its musical fantasy. Like Pan himself, a classic that will never grow old.

*__Peter Pan__ (Walt Disney, 76 min., ages 3–10) Disney classic where the quintessential fairy was born in Tinkerbell and every scene glows with magic. From the moment Peter exhorts "You

Can Fly!" right up to his dramatic pirate rescue of Wendy and the boys, the fantasy never lets up. Never Land is a wellspring of adventure, including one with unfortunately stereotypical "Injuns." Still, *Pan* is a near-perfect paean to young imaginations.

Pinocchio (Walt Disney, 87 min., ages 4–9) This Italian fairy tale of a puppet come-to-life is itself brought to beautiful life by Disney animators. Nearly every scene offers the Old World charm of scrolled detail and muted colors. And the Blue Fairy's appearances, though brief, are standards of magical grace. While its morality is laid with a heavy hand, *Pinocchio* remains a boyhood adventure well worth taking.

***Puss in Boots** (Hanna-Barbera, 30 min., ages 4–9) One of the Timeless Tales From Hallmark, this animation offers a comedic retelling of the Charles Perrault classic. The story itself is amusing, of course, as a cat single-handedly changes his master's fortunes. What makes this version special is the characterization of Puss as a supercilious strutter who sings about being the cat's meow. As with other Timeless Tales, the art, animation, and writing all contribute to a good story well told.

Puss in Boots (Rabbit Ears, 30 min., ages 4–8) From We All Have Tales, the French classic is animatedly read by Tracey Ullman, with Jean Luc Ponty's music lending added sparkle. The artwork is simply, charmingly drawn by French illustrator Pierre Le-Tan. And the story's natural humor is well brought out in this fine adaptation.

Rapunzel (CBS-Fox, 45 min., ages 4–10) Faerie Tale Theatre video starring series producer Shelley Duvall, Jeff Bridges, and Gena Rowlands. Contemporized with a script of humor and gender equality, *Rapunzel* remains a fairy tale in spirit. Enchantments for good and evil drive the story of the imprisoned maiden. With art design based on traditional illustrations, *Rapunzel* is a joy for the whole family.

***Rumpelstiltzkin** (Hanna-Barbera, 30 min., ages 3–9) This Timeless Tale From Hallmark wonderfully adapts the Brothers Grimm classic. Its animation is well drawn and written, with a

Rumpelstiltzkin who speaks in clever verse. And its original songs underscore important moments in the story. Add a heroine who's attractive but no great beauty and a different twist at the end, and you have a new classic.

Scholastic Blue Ribbon Storybook Video, Volume 2 (Warner, 26 min., ages 3–7) *The Bremen Town Musicians* from the Brothers Grimm opens this animated video hosted by a donkey puppet. The story is well told and includes a cute song about animal noises. *Harold*, based on the Crockett Johnson books, uses his magic purple crayon to draw wherever he'd like to go. When he gets hungry, he just draws nine pies that have to be finished by a drawn moose and porcupine. A beautifully simple idea that celebrates imagination and creative problem-solving.

Shelley Duvall's Tall Tales and Legends After the success of her Faerie Tale Theatre series, Shelley Duvall turned to American folk stories. Assembling fine casts and giving them scripts with a contemporary edge, she breathed new life into the likes of *Annie Oakley, Darlin' Clementine, Johnny Appleseed,* and in the Holidays: Halloween section you'll find *The Legend of Sleepy Hollow.*

Sign Me a Story (Random House, 30 min., ages 3–7) *Little Red Riding Hood* and *Goldilocks* are both acted and signed in this special video. First Linda Bove of Sesame Street teaches a few signs, such as "big" and "girl." Then she and her friends act out the story using the signs. Finally, she reviews the signs' meanings as they appear in the play. Narrated as well as signed, these two classics are also fun for hearing children, who'll learn a new language.

Sleeping Beauty (Walt Disney, 75 min., ages 3–10) Not one of Disney's best but nevertheless fine animation of a classic fairy tale. On the debit side are overstylized artwork, a weak main character, and songs that, excepting Tchaikovsky's title piece, are forgettable at best. On the positive side are the charming little fairy godmothers, their on-again-off-again magic, and the climactic finale battle with Maleficent. Worth renting and searching through the slower parts.

Snow Queen, The (Lightyear, 30 min., ages 4–10) A joint project of U.S. and Russian animators, this adaptation of Hans Christian Andersen's tale is based on the Susan Jeffers book. Sigourney Weaver gives a suitably haunting reading. And music from Jason Miles is as cool and magical as the story. When her playmate is kidnapped by the Snow Queen, young Gerda sets off to find and rescue him. With a little help from her friends, Gerda discovers that ultimately her "power is in her heart," with goodness and innocence her only weapons. Beautifully told and imaged, *The Snow Queen* is a different kind of fairy tale—more serious than most and featuring a strong young heroine.

***Snow White and the Seven Dwarfs** (CBS-Fox, 50 min., ages 4–10) Elizabeth McGovern is radiant in the title role and Vanessa Redgrave superb as the evil Queen in this Faerie Tale Theatre production. Add Vincent Price as the mirror, Rex Smith as the singing Prince, and accomplished dwarf actors, and you have a cast worthy of the classic. The excellence extends to a poetic script, art direction inspired by N. C. Wyeth, and truly magical scenes spun by Redgrave. All in all, a jewel of a video.

***Steadfast Tin Soldier, The** (Hanna-Barbera, 30 min., ages 3–10) This Timeless Tale from Hallmark updates a classic with fine animation, music, and bits of humor. When the villainous Jack-in-the-box sends him out the window and into adventure, the Tin Soldier remains steadfast by reciting principles from his soldier's handbook. Naturally, the Hans Christian Andersen story turns on the romance between the soldier and a toy ballerina. A tale that's not only timeless but ageless, appealing to young and old.

Stormalong (Rabbit Ears, 30 min., ages 5–10) From the American Heroes & Legends series comes this seafaring tall tale. John Candy narrates with appropriate bluster and a bit of a brogue. Stormalong, born of a sea storm, "was tying knots before rope was invented." When he gets too big for his ship, Stormalong heads inland, but not for long. He soon returns to Boston, where his friends have built him a clipper ship so big that "Congress granted her statehood." But can The Curser win a race against a steamship? Providing excellent complements to this John Henry-type story are music by NRBQ and brilliant pastels by Don Vanderbeek.

Tale of the Frog Prince, The (CBS-Fox, 50 min., ages 6–10) Faerie Tale Theatre gem starring Robin Williams and Teri Garr, and written and directed by Monty Python's Eric Idle. Williams, of course, is fantastic as the electronically-miniaturized frog. He's hysterical whether quipping with the Princess, sparring with his would-be chef, or doing table-top stand-up for the royal court. The dialogue gets a bit racy at times in this freewheeling adaptation. It retains a fairy-tale atmosphere, however, with set and costume design inspired by Maxfield Parrish.

*__Thumbelina__ (SVS, 30 min., ages 4–10) This Rabbit Ears Storybook Classic presents the Hans Christian Andersen tale with all the delicate charm of its heroine. Although not animated, the video sparkles with the crisp narration of Kelly McGillis, magical touches in Mark Isham's music, and entrancing artwork by David Johnson. Reminiscent of Cassatt, its graceful lines and glowing pastels favor this gem of a fairy tale.

*__Ugly Duckling, The__ (Rabbit Ears, 28 min., ages 3–10) Cher gives a surprisingly soft and effective reading in this Storybook Classic by Hans Christian Andersen. The slightly animated tale of pride and prejudice also benefits from original music by Patrick Ball and earth-toned art by Robert Van Nutt. The ending is beautiful, both figuratively in its moral *and* literally in a scene of dazzling white swans against a sunset sky and deep blue water. A treasure for the whole family.

Yeh-Shen (CBS-Fox, 25 min., ages 5–9) This Chinese Cinderella story predates the European version by a thousand years. An orphan lives with a cruel stepmother and stepsister. A golden fish, who befriends Yeh-Shen, promises her a better life. He delivers by becoming golden slippers that magically array her in finery for the festival. Nicely drawn and animated, this CBS Storybreak Video offers some interesting twists on a classic.

HOLIDAYS

Following are special videos for special times, guaranteed to put you in the appropriate holiday spirit. Think of them not only for kids' parties at home or in school, but for your own family celebrations. The videos in this section are secular presentations of the holidays.

VALENTINE'S DAY

Berenstain Bears and Cupid's Surprise, The (GoodTimes, 25 min., ages 3–6) Brother Bear tries to prepare for the upcoming hockey championship, but he's distracted by mysterious valentines from a secret admirer. Clues point to at least three suspects, with the ending truly a surprise. Subplot follows Papa's attempts to build an impressive valentine for Mama. With its funny scenes, a little mystery to solve, and the usual Bear rhymes, this is a fine Comic Valentine (the video's original title).

Family Circus Valentine (FHE, 53 min., ages 2–6) Positive, warm-hearted cartoon starring the Bill Keane characters. As the kids compete to make the biggest valentine, they make funny mistakes and learn the real meaning of love. Keane's childlike drawings are nicely animated, and a sweet song opens the tape with a smile.

Peanuts: Be My Valentine, Charlie Brown (Lee Mendelson, 25 min., ages 3–8) Lucy pines for Schroeder, Linus has a crush on his teacher, and Sally Brown one on him. But all Charlie Brown wants is a valentine from somebody, anybody. Briefcase in hand, he's the picture of optimism on Valentine's Day. Snoopy performs a cute, one-dog play with his "pawpets." And little absurdities pop up here and there, like valentines from Charles Schulz. A heartwarmer.

Peanuts: It's Your First Kiss, Charlie Brown (Lee Mendelson, 30 min., ages 5–9) Love and excitement are in the air as the school homecoming approaches. Just as you get past the absurdity of Charlie Brown's playing on the football team, you're hit with another. As the homecoming queen's escort, he must kiss her at the dance. Another Peanuts classic, with humor in every artful squiggle.

EASTER

Berenstain Bears' Easter Surprise, The (GoodTimes, 25 min., ages 3–6) Back when Brother was only a cub, Bear Country weather got stuck in winter. Brother can't wait for the Easter surprise Mama has promised, and it takes a visit from Papa to get the Easter Bunny going. Once again, it's clever rhyme time with the Berenstain Bears in this well-animated, amusing video.

Buttons and Rusty and the Easter Bunny (Summit, 25 min., ages 3–8) A "Chucklewood" Easter from former Disney animator Ed Love. Buttons and Rusty are cubs, a bear and fox respectively, who fall into innocent mischief during holidays (see also Halloween and Christmas entries). Here they try to get in the Easter act by borrowing various nest eggs to color. The eggs' mistaken return results in such comical scenes as a turtle trying to walk like a duck. The story is as cute as the characters, well animated and nonviolent.

Easter Bunny is Coming to Town, The (FHE, 50 min., ages 3–8) Rankin-Bass "animagic" special told and sung by Fred Astaire. The story takes place in Kidville, known to have the best eggs anywhere. When the children take on a pet bunny and learn of another sadder Kidville, Easter is born. Woven throughout the story are cute songs and clever origins of such traditions as jelly beans and hiding decorated eggs.

HALLOWEEN

***Boy Who Left Home to Find Out About the Shivers, The** (CBS-Fox, 50 min., ages 6–14) Faerie Tale Theatre production boasting such fright-night greats as Christopher Lee, David Warner, and as narrator Vincent Price. None can scare Peter MacNicol, however, who longs to experience the shivers. His three nights in a Transylvanian castle bring on live gargoyles, ghostly apparitions, a game of skull bowling, and, thanks to his unexpected reactions, a lot of laughs. An excellent production with fine acting, art direction, and horror-to-humor dynamic.

Buttons and Rusty: Which Witch Is Which? (Summit, 25 min., ages 2–6) Once again the little bear and fox cubs get into trouble trying to learn about holiday customs. Along the way, kids learn about trick-or-treat safety and wildlife rights. A *Fantasia*-like party scene of dancing goblins and jack-o'-lanterns is as scary as this animation gets, making it perfect for very young children.

Canterville Ghost, The (FHE, 49 min., ages 4–10) Like a turn-of-the-century *Beetlejuice*, this *Ghost* is at turns weird, funny, and just a shade scary. The Oscar Wilde tale may not be stunningly animated, but it's still highly entertaining. When an American family moves into the manor, the ghost of Lord Canterville finds them singularly unspookable. Their matter-of-fact treatment frustrates him until the sympathetic daughter aids his release to final rest. A great first ghost story.

Dicken's Ghost Stories (Celebrity, 60 min., ages 6–12) Animation of three selections from the pages of *The Pickwick Papers*. *The Ghost in the Wardrobe* tells the humorous tale of a spirit who mysteriously returns to the same closet each year. *The Mail*

Coach Ghosts menace a young traveler who defends fair maiden. And *The Goblin and the Gravedigger* features a Scrooge-like character who reforms after a fiery trip to hell. Concise and loaded with Dickensian atmosphere, all three appeal to kids.

***Donald's Scary Tales** (Walt Disney, 22 min., ages 3–10) Halloween triple feature of tricks and treats. In *Donald and the Gorilla,* his nephews don a gorilla costume even as the real thing escapes from the zoo. *Duck Pimples* finds Donald living in the radio programs he's "enjoying." And *Donald's Lucky Day* casts him as a messenger who has to deliver a ticking package by midnight. Opening with a snappy dance, his fortunes quickly fade as he goes into the night of Friday the 13th. Close encounters with superstitions show Donald in classic Duck form.

Frankenweenie (Walt Disney, 30 min., ages 6–12) Director Tim Burton's (*Batman, Beetlejuice*) first film, made when he was at Disney in 1984. Shot in black-and-white, à la horror classics, the unusual film stars Shelley Duvall, Daniel Stern, and a great dog in the title role. When he's brought back to life by his young owner, Frankenweenie terrorizes the neighborhood. It's not that he wants to, he's just a misshapen, misunderstood mutt. Like Burton's feature films, this short is funny and weird at the same time.

House of Dies Drear, The (Public Media, 110 min., ages 6–14) WonderWorks title starring Howard Rollins, with Kadeem Hardison in a small role. The big question is Who is haunting this house? The ghost of Dies Drear, former owner and abolitionist who helped slaves escape on the Underground Railroad? The ghost of a slave who lived in the property's caverns? Or Pluto, the house's devilish caretaker? Or maybe the Darrow family, who covet the property for themselves? Thomas Small and family encounter some scary stuff on the way to solving this mystery. A good and educational suspense yarn.

***Legend of Sleepy Hollow, The** (CBS-Fox, 50 min., ages 6–12) A bright one from Shelley Duvall's Tall Tales and Legends. Cast in the golden glow of firelight, this production makes the most of the early American ghost story. The aura of mystery holds throughout, even with comic touches, right up to its spooky climax. Ed Begley, Jr., turns in a fine performance as the gullible Ichabod Crane, Beverly D'Angelo plays Katrina, and Charles Durning her uncle and wry narrator.

*__Legend of Sleepy Hollow, The__ (SVS, 30 min., ages 4–10) Rabbit Ears Storybook Classic of romantic rivalry and ghost stories of the headless horseman. Glenn Close narrates this tale for "crisp autumn evenings" in suitably hushed tones. And illustrations by Robert Van Nutt recall the simple charm of early American art. Overall, the tone is eerie without being frightening—just right for the Washington Irving tale and for family viewing.

Peanuts: It's the Great Pumpkin, Charlie Brown (Lee Mendelson, 25 min., ages 3–7) Halloween special starring Linus as the sole believer in the Great Pumpkin. Let the others ready for trick-or-treating and Halloween parties; Linus awaits the great arising in the pumpkin patch of his dreams. It's an endearing story told with characteristic Peanuts charm. The look on Linus's face, of innocent faith despite others' skepticism, is enough to make a believer out of anybody.

*__What's Under My Bed and Other Creepy Stories__ (Children's Circle, 35 min., ges 4–9) Book adaptations include the iconographic *Georgie* by Robert Bright and three animations. James Stevenson's title story has a clever grandpa getting his young houseguests to calm themselves down at night. Animated by Michael Sporn with an Ernest Troost score, it's a wonderfully imaginative tale in jewel-toned watercolors. *The Three Robbers* by Tomi Ungerer find themselves reformed by a brave little girl. And *Teeny-Tiny and the Witch-Woman,* based on an old Turkish tale, is a good 'n spooky story with beasties in every corner. Teeny-Tiny, the youngest of three brothers, saves them all from the witch-woman with a bit of trickery. With richly-colored medieval art and haunting music and narration, *Teeny-Tiny* offers family-size Halloween fun.

CHRISTMAS

*__Babar and Father Christmas__ (Media, 30 min., ages 3–6) Fully animated adaptation of the Jean de Brunhoff book as narrated by his son Laurent. Faithful to the book's charms, the video's scenes are wonderfully composed and detailed, leading into a special world. The story, too, is engaging, as Father Christmas comes to the land of the elephants and gives King Babar a magical gift.

*__Baby-Sitters Club Special Christmas, The__ (Scholastic/Good-Times, 30 min., ages 6–14) This video is indeed special. Not only is it full of the sights and sounds of the season, its Christmas spirit springs from original sources. Central to the story is the girls' party for children in the hospital. Diabetic Stacey finds herself a patient there after sampling too many Christmas cookies. Fortunately, the viewer doesn't OD on "sugar" because the video is very natural and well written. At its end is an act of such generosity, only the jaded will remain unmoved.

*__Baby Songs Christmas__ (Golden, 26 min., ages 3–6) Like others in this fine line of live-action videos, *Christmas* offers very singable arrangements of both original and traditional tunes. And they're staged in clever, kid-appealing ways. In "The Wassail Song," carolers get waffles and fossils instead. And "We Wish You a Merry Christmas" animates the likes of Russian dolls and nutcrackers. Other clips include the multicultural "Peace on Earth," a funny class performance of "The Twelve Days of Christmas," and "Up on the Rooftop" doo-wop.

*__Barney and the Backyard Gang: Waiting for Santa__ (Lyons, 45 min., ages 2–6) Holiday offering of the popular sing-along series starring Barney, the purple dinosaur. When he suggests the gang befriend the new boy in town, they learn Derek is afraid Santa won't find him in his new home. So Barney whisks everyone to the North Pole for a magical musical fantasy. Best cuts are "S-A-N-T-A" sung to the tune of B-I-N-G-O, "Jingle Bells," and "Let's All Do a Little Tapping" staged in Santa's workshop. Nice wrap-up is Barney's reading of *The Night Before Christmas.*

***Berenstain Bears' Christmas Tree** (GoodTimes, 25 min., ages 3–8) It's Christmas in Bear Country, and all that the family "bear-loom" ornaments need is a tree. Papa decides it can't be just any tree, but in his search he learns that trees are homes for creatures, too. The solution is happy, and its meaning delivered with a light touch. Inventive animation and cute songs add to the Christmas spirit.

Bluetoes, the Christmas Elf (FHE, 27 min., ages 3–7) Canadian animation that tells a sweet and funny story. In a new twist, toys grow on trees high in the mountains where elves pick and send them down a chute. When a "small one" accidentally breaks the chute, he looks for a way to redeem himself. His solution not only begins a tradition but earns him the name of Bluetoes.

Buttons and Rusty: A Special Christmas (Summit, 25 min., ages 3–8) The little bear and fox cubs hop on the Christmas-tree train in this animated adventure. When they arrive in the city, they're surprised at the "tall cabins" and strange city "animals" with wheels and sirens. Finding their way to a department store, Buttons and Rusty amuse themselves—and others—in a toy window display. An original and nicely animated holiday story.

CareBears Nutcracker Suite (GoodTimes, 60 min., ages 3–8) The CareBears cut down on their sweets long enough to make this Christmas story. Moving between Care-a-lot and Toyland, they help a girl named Anna and her Nutcracker doll fend off the rat king and an evil vizier. Liberties are taken with the Hoffman tale, but it's a good story centering on the Sugar Plum Fairy and her magic ring. And the animation is attractive, with musical numbers well integrated. A nice intro for little ones.

Child's Christmas in Wales, A (Vestron/FHE, 55 min., ages 6–10) This video adapts the Dylan Thomas poem with appropriate and unhurried grace. His remembrance of Christmas past unfolds in conversation between a boy and his grandfather, nicely played by Denholm Elliot. And what a memory it is—told in vignettes that unabashedly reveal the charm of times gone by.

The language is, of course, beautiful, with visuals to match. Curl up with it one quiet evening over the holidays.

***Disney Christmas Gift, A** (Walt Disney, 47 min., ages 3–10) Seasonal scenes from animated features are teamed with Christmas cartoons from the Disney vaults. *Pluto's Christmas Tree* comes with the rascally Chip 'n Dale to wreck the halls of Mickey's home. *The Clock Watcher* is Donald Duck, assembly-line toy wrapper who takes on a punchy Jack-in-the-box. And *The Night Before Christmas* makes a grand finale. Animated by Ub Iwerks, W. Clement Moore's poem is beautifully drawn and set to music. The right jolly old elf's toys parade out of their sack, decorate the tree, and then dance around it to a clown band. A holiday gem worth seeking out.

***Dr. Seuss's How the Grinch Stole Christmas** (MGM-UA, 26 min., ages 3–10) Out to out-Scrooge Scrooge, the Grinch plots to steal Christmas from Whoville. And he nearly succeeds, despite the mismatch of his packed sleigh and pathetically small dog of a "reindeer." Other sights of the season include the littlest Whos puzzled at the Grinch Santa's *taking* toys and then his heart growing by leaps at the end. The sounds of the season are memorable songs amid Ted Geisel's inimitable verse. A Christmas classic perfectly read by Boris Karloff.

Frosty the Snowman (FHE, 30 min., ages 3–7) Jimmy Durante narrates and sings this animated story built around the classic song. Frosty comes to life thanks to a discarded magic hat, only now the magician wants it back. So Frosty, with the rabbit who came with the hat and a little girl, must elude the magician long enough to get to the North Pole. Well-drawn and animated, *Frosty* finishes with a little Christmas magic from Santa Claus.

Garfield Christmas Special (CBS-Fox, 25 min., ages 4–10) Even as Jon revels in Christmas spirit where *everyone* has a smile, Garfield proves the frowning exception. This sentimental-cynical contrast carries over to their holiday visit at Jon's family farm. But then the fat cat finds a soul mate in Grandma and comes around. Lest things get too sweet, the cartoon ends with the traditional Eve story, *How Binky the Clown Saved Christmas.*

Gingham Dog and the Calico Cat, The (Rabbit Ears, 25 min., ages 3–8) A sweet and loving Christmas tale that begins as anything but sweet. Two competitive elves fashion the dog and cat toys without love. So it's not until they fall out of Santa's sleigh and into the woods that they learn to help and depend on each other. The toys' adventures are intercut with the story of their eventual owners, a brother and sister who fight like you-know-whats. Grammy artist Amy Grant narrates, with wonderful pastel art and Chet Atkins music accompaniment.

Little People Christmas Fun (R&G, 30 min., ages 2–6) Animated characters from the Fisher-Price toy line celebrate Christmas with a trip to Grandma and Grandpa's. When Penny runs out of money before buying Timmy's present, she frets until Grandpa reminds her the best gift is from the heart and not a store. Charmingly drawn and pleasantly paced, *Little People* has family Christmas spirit.

*__Madeline's Christmas__ (Golden, 23 min., ages 3–6) Narrated by Christopher Plummer, this animated story opens with our heroine sledding and skiing and singing brightly "I'm Madeline." She needs her good spirits, too, for it looks as though Christmas will be spoiled by a snowstorm and illness at the girls' school. With the help of magical Marie, however, Madeline manages the best Christmas ever. Ludwig Bemelmans's poetry is perfectly complemented by the original music. And his art is beautifully animated, from the opening Paris snowfall to a Christmas party finale.

Mickey's Christmas Carol (Walt Disney, 25 min., ages 3–8) A fine adaptation of the Dickens classic starring Mickey, in his first role in thirty years. Simplified for young viewers and yet not trivialized, *Carol* stars all the familiar characters. In a bit of typecasting, Uncle Scrooge plays his namesake, Mickey is Bob Cratchit, with supporting roles played by Donald Duck, Goofy, and

Jiminy Cricket. A lighthearted intro to the holiday's most famous story.

Mister Magoo's Christmas Carol (Paramount, 52 min., ages 4–10) Imaginative rendition of the Dickens classic casting Magoo as Scrooge. The usually lovable myopic does a credible job in the role, too, thanks to the voice talents of Jim Backus. Add a fine and funny script, with bright score by Jule Styne and Bob Merrill, and you have a great intro to a classic.

*__Nutcracker, The__ (Kultur, 100 min., ages 6–14) Stunning performance of the Bolshoi Ballet, the company which commissioned Tchaikovsky to score *The Nutcracker* over 100 years ago. Choreographed by Yuri Grigorovich, the ballet stars Yekaterina Maximova and Vladimir Vasiliev. Its staging is masterful, the music sparkling, and choreography inventive. Not a false note is struck in this production that draws both tingles and tears. A Christmas treasure.

Santa Claus Is Coming to Town (FHE, 53 min., ages 3–6) Fred Astaire tells and sings this "animagic" story of how Santa Claus and his traditions came to be. As an abandoned baby, he is taken in by elves, the Kringles, who name him Kris. They make toys but can't transport them, and the kids of Sombertown want toys but aren't allowed to have them. This, of course, is where Kris Kringle steps in and solves everybody's problems with Christmas traditions. The story neatly explains such things as why Santa comes down chimneys and leaves presents in stockings.

SantaBear's First Christmas (Vestron/FHE, 25 min., ages 3–7) Animated story, narrated by Kelly McGillis, of how SantaBear earned his name. Separated from his parents, the baby bear is taken in by a young girl. Marie teaches him to bake and dance and read. When her grandfather becomes ill and only Santa Claus can help, the little bear has to deliver all his presents. While only slightly animated, the video's artwork is as warm and beautiful as the story.

SantaBear's High-Flying Adventure (Vestron/FHE, 23 min., ages 3–7) When the evil Bully bear tries to discredit Santa by stealing his toys, it's SantaBear to the rescue. Kelly McGillis narrates with help from John Malkovich (Santa Claus) and Bobby McFerrin (SantaBear). Art and animation by Michael Sporn dazzle in scenes of night flight and sparkling ice igloos. In the end, Santa tells the two bears apart by asking the true meaning of Christmas.

Sesame Street Special: Christmas Eve on Sesame Street (Random House, 60 min., ages 3–6) 1978 television special that's a potpourri of holiday skits and songs. It opens with a question posed by Oscar to Big Bird: How does Santa Claus, who is built like a dump truck, get down all those skinny little chimneys? By tape's end, Big Bird has solved the riddle and discovered the true miracle of Christmas, the spirit of loving and sharing. Along the way, Kermit reports on theories involving rubber chimneys or very long arms, kids give their own ideas, and Bert and Ernie exchange presents with a nod to O. Henry's *Gift of the Magi.*

Simpsons' Christmas Special, The (CBS-Fox, 30 min., ages 4–12) Yes, Virginia, the Simpsons do celebrate Christmas, just not in the usual ways. Lisa plays Santa of the South Seas in the school pageant, Homer takes a stab at being a department-store Santa, and Bart gets a tattoo. At once touching and taunting, and of course rib-tickling, this is Christmas as only the Simpsons could create it.

*__Tailor of Gloucester, The__ (SVS, 30 min., ages 4–12) Rabbit Ears Storybook Classic based on a Beatrix Potter tale. Read with a loving lilt by Meryl Streep, the story is also favored with the soft pastels of David Jorgensen and original music by Paddy Maloney and the Chieftains. It's Christmas Eve, and as a poor tailor rushes to finish the mayor's wedding suit, he runs out of cherry thread. His lament of "no more twist!" is heard by his mischievous cat as well as some nimble-fingered mice. A *Tailor* to weave a spell of Christmas magic over the entire family.

*__Walt Disney Christmas, A__ (Walt Disney, 46 min., ages 3–10) A classic collection of holiday cartoons. In the best, Mickey and Minnie Mouse cavort *On Ice,* and there are the usual shenanigans in *Donald's Snow Fight.* A real Christmas treat is the 1932 *Silly Symphony, Santa's Workshop.* Santa and his elves bustle with Christmas Eve activity, as toys are cleverly assembled and set in motion by a rollicking airplane. Classic Disney for the season.

Wee Sing: The Best Christmas Ever (Price/Stern/Sloan, 60 min., ages 3–8) Based on the best-selling book and cassette *Wee Sing for Christmas,* this live-action video stages twenty holiday songs and finger plays. They're sprinkled throughout the story of Poofer, an elf who brings a family to Santa's workshop. The songs, both original and traditional, are nicely arranged and sung for children to sing along. And fine art direction adds to the holiday spirit.

INSTRUCTION

If a picture is worth a thousand words, moving pictures are invaluable, especially when they are instructional. Learning how to draw, hit a baseball, or perform magic is all the more effective when you can see how a master does it . . . *and* can review the lesson as often as you'd like. Following are the best instructional videos for kids—those that make learning fun—organized into categories of art, dance and exercise, sports, and miscellaneous.

ART

Be a Cartoonist (Mid-Com, 60 min., ages 5–10) Professional Alan Silberberg takes numbers and letters and turns them into amazingly effective cartoon characters. Like magic, he says, cartooning is as easy as 1, 2, 3—and then he draws a creature with the numerals. Silberberg is an excellent teacher who explains things well, generates ideas, and makes learning fun. A creative and funny video.

Draw Squad: Moonbot's Birthday Party (Bridgestone Group, 40 min., ages 6–10) Instructor Mark Kistler, of National Public Television's "The Secret City," may overpunctuate his lessons with corny songs, dances, and jokes. But he does give solid, kid-appropriate art instruction embedded in good little stories. In this third, and best, volume, the story concerns a rocket launch to the moon for a birthday party. Other volumes include *Escape of the Twinkies* and *Pigusis Goes on a Diet.* See also *The New Secret City Adventures* and *The Seven Magic Words of Drawing*.

How to Create a Comic Book (Stabur, 50 min., ages 10 and up) Aspiring comic book artists can learn from the best in this video starring Todd McFarlane and Rob Liefeld. Creating a cooperative work, they give tips in drawing, page layout, and how to interest editors in your work. They don't, however, talk much about story writing or character creation. Still, this is a great art video which benefits from Liefeld and McFarlane's humorous banter and a house call from Jim Lee to teach anatomy-based drawing.

***Look What I Made: Paper Playthings and Gifts** (Pacific Arts, 45 min., ages 6–10) Fine instruction in seven projects ranging from the simplest flowers to magical origami to a Mexican piñata. The video is colorful and well produced, using miniraps to list materials. Plus it stars the extraordinary Amy Purcell, whose sense of humor helps her and your child cope with mistakes.

My Fun Pack series (Morris, 30 min. each, ages 6–10) These art-craft videos for kids are chock-full of original, workable ideas. What instructor Julie Abowitt lacks in dynamism, she more than makes up for in creativity and teaching ability. Bits of art education are deftly woven into project demonstrations. Because some need adult supervision, you should watch with your child the first time. Here are the best:

Fun With Clay Using both pinch and coil methods, Abowitt shows how to make clay tiles, jewelry, buttons, beads, and animals. Also included are care of the medium and workspace as well as sculptural tips, such as looking at a piece from all angles.

Paint Without a Brush Explores such techniques as wet chalk and dribble painting. All are fun, easy, and allow plenty of creative expression. As she paints with a cotton swab, Abowitt talks about the fun of the process and its similarity to Impressionism. She also takes childlike delight in art's little "moments of surprise."

Paper Play "Art is about discoveries," Abowitt shows as she works her way through a half-dozen tissue paper projects. While making collages, vases, wrapping paper, and "stained glass," she shares such art concepts as composition and hot and cool colors. A fine tape that points you in your own creative direction.

Vegetable Print Shop After a discussion of what "print" means, Abowitt makes prints with potatoes, apples, pears, even artichokes. Good tips include getting an adult to do the cutting and using food coloring so you can still eat the fruit afterwards. The resultant art has impressive, natural design.

New Secret City Adventures: Volume 1, The (Mark Kistler, 30 min., ages 5–12) Commander Mark praises books in this adventure. His drawings include readers and a library that looks like a book. And his instruction is peppered with minisongs, bad jokes, and good talk about imagination. Pencil Power is more than following his direction, it's adding your own ideas to create a style. Successful and entertaining video that ends with an addition to the galactic Neverending Drawing.

*__Seven Magic Words of Drawing, The__ (Mark Kistler, 60 min., ages 6–14) "Anybody can draw," declares art instructor Mark Kistler. Watching this video, you believe him. Star of PBS's "The Secret City," Kistler gives seven tips for creating the illusion of 3-D in drawings. Each is taught and then incorporated into a picture with previous magic words. It's a great system, translating sophisticated art concepts to kid-level tricks, and Kistler's animated style holds kids' attention so they can learn it.

*__Squiggles, Dots, and Lines__ (Kidvidz, 30 min., ages 6–14) This descriptive title represents the drawing alphabet of Caldecott-winning artist Ed Emberley. His system is so simple and yet endlessly variable that it allows for quick success *and* long-term creativity. And the video itself is entertaining as computer animation brings his drawings to life and transforms them. A celebration of creativity that informs and amuses children.

Wow, You're a Cartoonist! (Lorimar, 30 min., ages 6–10) Jim Henson Play-Along Video hosted by Kermit the Frog. The first section is Faces, where a beret-ed Muppet shows how to draw a cat, and then such famous Muppets as Miss Piggy and Fozzie

Bear. The Figures section has a teenage artist instruct in adding bodies to the faces. Next up is Kermit with animal sketches and a bright song about doodling. And finally a segment on cartoon panels and flip books shows creative ways to use your drawings.

DANCE AND EXERCISE

***Ballet Class for Beginners** (Kultur, 40 min., ages 8–14) Instructor David Howard has coached the likes of Gelsey Kirkland, Natalia Makarova, and Mary Tyler Moore. This video is like a private lesson, moving from posture to positions to exercises and then minidances. Throughout, Howard narrates while a student demonstrates. His coaching—exacting, thorough, and inspiring—should provide an excellent complement to ballet lessons. See also *Tap Dancing for Beginners.*

***Creative Dance for Preschoolers** (Butterfly, 60 min., ages 3–6) Instructor Kathy Blake uses children's eagerness to move with music to introduce them to dance. Her playful attitude comes through even as she teaches such important concepts as remembering a pattern and counting beats. Typical of her expert, age-appropriate direction is suggesting that your body make shapes to match the music. One of six in the children's series *Let's Learn How to Dance.* See also *Jazz Dance for Kids.*

I Can Dance (JCI, 30 min., ages 7–10) An introduction to ballet with instructor Debra Maxwell. Opening with young Emily's dream to be a ballerina, the video moves on the five basic positions, barre work, and a discussion of the art. Ending with centre work, the video makes an excellent trial and beginner practice tape.

Jazz Dance for Kids (Butterfly, 60 min., ages 6–11) Excellent beginner tape from instructor Kathy Blake. She's great at breaking down a dance into simple steps. After teaching the ball-change, she drills it and then incorporates it into a dance. By tape's end, Blake leads you through a dance slowly, with beat accompaniment and calling the steps. Then it's faster and with music for an upbeat finale. See also *Creative Dance for Preschoolers.*

Tap Dancing for Beginners (Kultur, 30 min., ages 8–14) Instructor Henry LeTang is an expert teacher who guides a student from the basic shuffle-step to complicated combos. His progressive method begins with a single step, broken down into its parts. After a good drill, the step is done to music and then added to other steps in a routine. A comprehensive, quick-moving lesson for beginner to intermediate tappers. See also *Ballet Class for Beginners.*

Tip Top With Suzy Prudden, Volumes 1 and 2 (Warner, 48 min. each, ages 3–6 and 7 and older) Peppy workouts for 3 to 6 year-olds and children 7 and older respectively. Prudden may be a bit too "tip top," but her exercises are creative and age-appropriate. She draws preshoolers into movement with pretend play. Older children engage in sportslike action. Because each video has six miniworkouts, they can be done a few at a time.

***Workout With Daddy and Me** (FHE, 30 min. ages 3–6) A better title might have been *Play*out. The four sets of dads and kids on this tape have so much fun, you want to jump right in. Against a backdrop of ever-changing, colorful computer art, the pairs play horsey, ride bikes in the air, and pretend to be anything from giants to flagpoles. Naturally this involves exercise and just as naturally it involves father and child together.

***Workout With Mommy and Me** (FHE, 30 min., ages 3–6) Dance educator Barbara Davis leads her own girls, and two other mom-kid sets, in creative movement that exercises the imagination as well as the body. Against a colorful, changing backdrop, they pretend to be growing flowers, butterflies, rockets, sandwiches even. The music is upbeat and Davis an engaging "storyteller" throughout the movement tale. A work*out*standing to give you and your child something different to do together.

SPORTS

***BASEBALL: Mickey Mantle's Baseball Tips** (CBS-Fox, 60 min., ages 6–12) The legendary Yankee's relaxed approach belies the wealth of education in this tape. After a warm-up and equipment talk, Mantle and friends instruct in batting, fielding, and pitching. All the basics, as well as special hints, are delivered as smoothly as a slider. Not only kids but newly-drafted coaches should find these tips invaluable.

BASEBALL: Play Ball With Reggie Jackson (ESPN, 30 min., ages 6–10) With a little help from his friends, Jackson coaches boys and girls in pitching, batting, base running, and fielding. He sets a natural, relaxed tone to the video which emphasizes the right way to do things. Showing very specific body motions for each action, Jackson also suggests drills that are actually fun. His friends include Andre Dawson, Ozzie Smith, and Rick Sutcliff.

BASKETBALL: Dr. J.'s Basketball Stuff (CBS-Fox, 60 min., ages 8–14) After highlights of Julius Erving's career, "The Doctor" conducts a clinic in the sport. Covered are the fundamentals in stance, defense, rebounding, passing, ball handling, shooting, and "the moves." In the House Call section, you put it all together and play defense against Dr. J. himself. Jam-packed with solid instruction for the right basketball stuff.

BASKETBALL: Put Magic in Your Game (CBS-Fox, 45 min., ages 6–14) Ever a generous player, Earvin "Magic" Johnson gives away some of his secrets in this video. After showing highlights of each, Magic suggests drills in dribbling, passing, rebounding, shooting, and defense. He also gives a magic tip for each part of the game. Keeping your head high and the ball low, for example, helps in dribbling. And when shooting bank shots, says Magic, give the backboard a name and "kiss" it with the ball.

***BASKETBALL: Teaching Kids Basketball With John Wooden** (ESPN, 75 min., ages 8–14) A superb coaching tape from a superb coach. UCLA legend Wooden does everything right, from teaching the fundamentals to giving the whys behind the whats to emphasizing team attitude and pride over winning. In the Shooting section, there are ten minutes of instruction on what to do before the ball ever leaves your hands. Also covered are Passing, Dribbling, Rebounding, and Defense. An invaluable aid to young players and their coaches.

FOOTBALL: How to Play Winning Football (Wood Knapp, 60 min., ages 10–14) Superbowl-winning Jim McMahon hosts, along with fellow quarterback Sonny Jurgensen and coach Tom Landry. In the opening section on Mental Preparation, each shares a tip—McMahon on visualizing your goals, Jurgensen on knowing your opponent, and Landry on team discipline. Then a section on Conditioning shows cardiovascular, sport-related exercises. Finally, there's good, detailed instruction in such skills as centering, passing and ball handling, receiving, and kicking. A good video for showing all the time and hard work behind the glory.

GYMNASTICS: Gymnastics Fun With Bela Karolyi (FHE, 60 min., ages 6–12) Coach to Olympic gold medalists Nadia Comaneci and Mary Lou Retton, Karolyi shares his teaching talents on video. He's an excellent and enthusiastic instructor, moving from fundamentals to breathtaking Olympian routines. Best segments show the moves in slo-mo, revealing not only their grace but step-by-step accomplishment.

SOCCER: The Graduated Soccer Method, Volumes 1, 2, and 3 (Wood Knapp, 30 min. each, ages 8–14) Excellent series of videos moving from *Building a Relationship With Your Soccer Ball* (Vol. 1) to *Developing Fast Feet/Shielding the Ball* (Vol. 2) to *Taking on Your Opponent* (Vol. 3). Instructor Gary McKinley, youth and collegiate coach, does voice-overs as kids demonstrate techniques. What "graduated" means is that each move is first done slowly, then faster, and then in relationship to another player. Very detailed and motivational videos.

TENNIS: Virginia Wade's Class (SVS, 67 min., ages 8–14) Like other pros, Wade stresses the practice and skill behind what appears simple and natural. Her advice is organized into sections, such as Forehand, Backhand, Net Play, and Serve. Each is a detailed clinic of teaching, slo-mo demonstrations, and a checklist of common errors. Covering a lot in a short time and giving the whys behind the techniques, Wade indeed has class.

MISCELLANEOUS

Be a Juggler (Mid-Com, 40 min., ages 6–10) Instructor Mike Vondruska teaches the four-thousand-year-old art step-by-step. Beginning with lightweight scarves included with the video, he then moves up to "jugaloons." These balloons, also included, are filled by you with salt or sand to make excellent juggling items. A good beginner-through-intermediate video that offers a variety of juggling styles.

Be a Magician (Mid-Com, 60 min., ages 8–14) Pro Martin Preston reveals that the true trick of magic is making the impossible seem real. So in addition to five classic tricks, he gives hints in performing them. In fact, he does the tricks in front of a mirror so you can see both what he does and what the audience sees. Some tricks are easy, others require sleight of hand that will take some practice. A good advanced video, complete with props, for the serious enthusiast.

Frisbee Disc Video, The (Wood Knapp, 35 min., ages 6–14) All you ever wanted to know about Frisbees—their history, physics, and moves both and simple and tricky. "Crazy John" Brooks, a world champion, teaches basic and advanced techniques for play, competition, and canine catches. And he offers tips for skipping the Frisbee, rolling it in a "golf" game, and maintaining MTA, (Maximum Time Aloft). A video that's fun to watch *and* learn from.

Hey, What About Me? (Kidvidz, 25 min., ages 3–6) If you have a new baby in the family, this is just what older siblings may be feeling. And it's the feelings of preschoolers who now share

Mom and Dad that are freely expressed in this live-action video. Youngsters speak for themselves, and the kid-to-kid message is that a new baby means change—but a lot stays the same, and some of the changes are good. Included are songs and games for preschoolers to play with the new baby.

Hey, You're as Funny as Fozzie Bear (Lorimar, 30 min., ages 6–10) Jim Hensen Play-Along Video is a primer of stand-up comedy. After instruction in such kiddie classics as knock-knock and elephant jokes, Fozzie invites your child to join him on stage. "The Big Show" at the end also includes magic. So let your child watch first and then perform with Fozzie for you.

Home Alone (Media, 30 min., ages 6–10) Malcolm-Jamal Warner hosts this special-interest video for latchkey kids. Well-produced and flecked with humor, this Home Alone is as entertaining as it is instructive. Warner demonstrates the do's and don'ts with a little help from his young friends. Even a child who's only occasionally home alone would find this tape a valuable reference.

Kids Get Cooking: The Egg (Kidvidz, 30 min., ages 4–9) A celebration of the "incredible edible" combining music, comedy, science, and art. Real kids tell each other corny—make that eggy—jokes, experiment with eggs, and cook hard-boiled and muffin-tin eggs. Leading the way are Herb and Bea, diner-owning puppets from the creator of Alf. Eggs-traordinary fun.

Let's Get a Move On! (Kidvidz, 30 min., age 3–10) Eight million children move each year. A major life event, its emotional aspects are often overlooked in the face of its busy-ness. So this video offers not only practical tips, on packing for example, but advice on how to cope with your family's feelings. Like other Kidvidz tapes, this one stars kids themselves in four segments: Getting the News, Getting Ready, Getting the Move On, and Getting Settled.

*__Rainy Day Magic Show, The__ (ConsumerVision, 20 min., ages 6–10) Hocus pocus, magic is the focus of this excellent video. Magician Mark Mazzarella leads viewers through eight tricks, all of which are sure-fire and easy to do. And no special props are needed, just things like rubber bands, paper clips, newspapers, and eggs. Each trick is shown three times—first performed, then done in slo-mo with instructive voice-over, and finally with step-by-step written directions. Mazzarella, a fine and pleasant teacher, also gives performance tips for a polished act.

Table Manners for Everyday Use (LandyVision, 40 min., ages 6–14) If you've ever been appalled at the sight of your children eating, this is the video for you. It's very thorough, with instruction in table setting, sitting, eating specific foods, and ending a meal. It's also a lot of fun, flecked with oldie film clips and scenes of do's and don'ts. Except for a minor contradiction about elbows on the table, *Manners* gives excellent demonstrations, descriptions, and whys and wherefores of the etiquette we could use three times a day.

Who? Me? Juggle? (Start Right, 30 min., ages 8–14) Imaginatively staged video celebrating a "unique human activity that improves timing and concentration." Host Dan Menendez and colleagues instruct in three- and four-ball juggling, with tips on establishing rhythm and on correcting common errors. A good intro that covers the basics and shows pros at their best.

***Yo-Yo Man** (Wood Knapp, 30 min., ages 6–14) Tommy Smothers stars in this instructional video that's as entertaining as the genre gets. It's funny, beautifully shot, and well organized—moving from the yo-yo's history to basic tricks and then an "advanced state of yo." A classy tape on a classic toy.

MUSIC

Most of the videos reviewed here are sing-alongs, a real staple in kidvid. But there are also concert, instructional, and educational videos designed to introduce children to the magic that is music.

Baby Songs series (Media except where noted, 30 min. each, ages 2–6) Take children's music veteran Hap Palmer, connect him with ace video producers who also happen to be moms, and you have the first music videos for kids. *Baby Songs* was also the first video without a licensed character to make the kidvid charts. And justifiably—because when it comes to preschool sing-alongs, nobody does it better. The songs are very kid-friendly, and the videos star preschoolers themselves at play and "work." Here are the best bets, all of which include a lyrics booklet.

* **Baby Songs 1** Premiere video in the line presents ten music videos for the younger set. Its songs are playful and often instructive, as in "My Mommy Comes Back," about a new preschooler, and "Share," with its suggestion to share toys for more fun with friends. Also outstanding is "Piggy Toes," the cutest celebration of these unsung wigglers as you're likely to see.

Baby Songs 2: More Baby Songs More well-sung, mid-tempo tunes for tots. With songs about learning to walk, sitting in a high chair, tickling, and playing horsey, this is the most truly baby-oriented video. The songs are nicely balanced, too, between upbeat rockers ("Wild and Woolly") and gentle ballads ("My Baby").

* **Baby Songs 4: Even More Baby Songs** Once again, singer-composer Hap Palmer takes the stuff of early childhood—teddy bears and finger foods, rag dolls and bubble baths—and sets it to

music of instant appeal. Visuals, mainly live-action with bits of animation, are well chosen and edited to keep the tunes fresh through many verses.

Baby Songs 5: John Lithgow's Kid-Size Concert Who'd have guessed? Acclaimed actor John Lithgow is loaded with musical talent as well. With acoustic guitar and a young audience, Lithgow serves up seven tunes, the best of which are his own compositions. Silly faces and sounds find their audience, as does the chance to sing along at times.

Baby Songs 7: Christmas—see Holidays: Christmas.

* **Baby Songs 8: Follow-Along Songs** (Golden) As usual in this hit series, the original tunes are catchy and upbeat without being frenetic. And they show real kids at play in every which way—make-believe, physical, musical, and games. This is the most interactive *Baby Songs* yet, with its "Bean Bag Alphabet Rag" game and the easy-to-make rhythm instruments of "Homemade Band." Kids love watching this video and so will you, if you love watching kids.

Barney and the Backyard Gang series (Lyons Group, 30-50 min., ages 2–8) If your child has yet to be introduced to Barney, have I got a treat for you. The big purple dinosaur, who leads children in song in this live-action series, is especially loved by the very young. *You* may not get it; Barney veers far from adult tastes. But children are enchanted and can't seem to get enough of him. His best releases, created by a pair of enterprising moms, are . . .

* **Barney Goes to School** When young Tina wishes school could be every day, Barney whisks her to a friendly preschool/kindergarten room. All the classic activity centers are there, as is the Backyard Gang to explore a typical school day. Well-conceived and produced, the video includes over a dozen songs and rhymes, like "The Teddy Bear Rap." Could be a big help in acclimating your little one to the idea of school.

Barney in Concert A video treat for Barney fans everywhere. Those in the Dallas area were lucky enough to catch him live in concert with the Backyard Gang. Both the performance and videotaping are well produced. And there are plenty of familiar songs to get the family audience up and singing. Highlights among the seventeen tunes are "We Are Barney and the Backyard Gang," "Grandpa's Farm," and "Mr. Knickerbocker," debuting Barney's new friend Baby Bop.

* **Barney's Birthday** Another well-conceived sing-along, set at Barney's birthday party. The kids check how much he's grown and Barney dances to "Growing." After Mexican and Filipino birthday songs, there's "Frosting the Cake," "Everyone is Special," and, of course, "Happy Birthday." Bright with song and dance, and with a personal message from Barney, *Birthday* makes a festive video for any happy occasion.

Barney's Campfire Sing-Along Barney and the Gang sing summer camp favorites in an attractive, wooded setting. From a picnic "Where the Ants Go Marching" to hiking with "The Little Turtle," they end up around the campfire, crooning such classics as "Sarasponda" and "Kookaburra." Over a dozen songs are well sung and staged for little ones to chime in.

Rock With Barney The Backyard Gang visits a movie studio, full of sets and props for staging over a dozen songs. Some are eco conscious ("Protect Our Earth"), others are borrowed from Raffi ("Apples and Bananas," "Down by the Bay"). All are imaginatively staged and singably arranged, with that undeniable Barney magic.

Waiting For Santa—see Holidays: Christmas

***Bill Harley: Who Made This Mess?** (A&M, 55 min., ages 6–10) He's a singer. He's a storyteller. He's two children's entertainers in one. Bill Harley's first video reflects his commitment to both art forms. Basically a filmed concert of half a dozen songs, they're broken up by an inventive story about sneakers. Bright-eyed and T-shirted, Harley looks every bit his kid persona as he sings the funny "You're in Trouble," "Cool in School," and the sweet acoustic solo, "Moon and Me." Whether singing or telling tales, Harley has clear rapport with kids and their parents.

Don Cooper's Musical Games (Wood Knapp, 30 min., ages 3–6) Nine original, activity-oriented songs encourage preschoolers to get up and move. "Do the Bug" shows little ones dancing with sheets for butterfly wings. And "Jungle Jamboree" offers a rap of imagining being different animals. All the songs are cute and well sung in this video which is also very well produced.

Don Cooper's Songs of the Wild West (Wood Knapp, 30 min., ages 3–8) Fashioning original songs with an Old West flair, the singer-songwriter looks to pioneers, cowboys, and Native Americans for inspiration. Best cuts are "Trapper Pete," about a Davy Crockett-type character, "Potatoes and Beans," lamenting a chuck wagon cook's limited menu, and "Dance of the Indian Nations," where dances' symbolic meanings are shown in soaring eagles and mighty buffalo.

Ella Jenkins: For the Family (Smithsonian Folkways, 30 min., ages 3–7) Like her friend Fred Rogers, Jenkins made this video talking directly into the camera and therefore to the viewer. Whether singing or storytelling, her rich, warm voice makes this one-on-one format work. Best cuts are the introductory "You'll Sing a Song," "Rhythms on a Rhythm Box," the chanted "Tah-Boo," and "Mary Mack." Well, Mary Mack*s* actually, as Jenkins talks about and performs several variations on the theme. An irresistible sing-along.

*__Ella Jenkins Live at the Smithsonian!__ (Smithsonian Folkways, 30 min., ages 3–7) Nobody doesn't like Ella Jenkins. Her charisma with children and adults comes through even on video. This intimate concert with about a dozen preschoolers shows them thoroughly entranced, laughing and loving her special brand of music. Opening with "Hello" and its international variations, she performs a dozen favorites, including "Did You Feed My Cow?," "And One and Two," and "Miss Mary Mack." Sharing great rhythms and fun melodies, Jenkins brings on a most joyful noise.

Frank Cappelli: All Aboard the Train (A&M, 30 min., ages 2–6) Nine music videos culled from Cappelli's Pittsburgh television program. Best cuts include "Rat-a-Tat-Tat" with its young percussionists, the rocker "Washing My Face," and "The Fruit Market," where Cappelli recalls Chico Marx in singing the praises of fruit. An excellent singer and songwriter, Cappelli takes obvious joy in his melodies and in the children he's educating.

Good Morning Sunshine (Golden Glow, 30 min., ages 2–6) Lives up to its billing as "songs for a day full of wonder." Singers Patti Dallas and Laura Baron give traditional children's songs a Renaissance spirit with dulcimers, recorders, and drums. Children are the real stars of the video as they laugh, skip, play, and dance to the music. Pretty and peaceful, "Sunshine" would make a nice wake-up video after naps, too.

***Joe's First Video** (ShadowPlay, 30 min., ages 3–10) Joe Scruggs is a children's musician par excellence. His bright, funny, finger-snappin' music is given visuals to match in this well-produced video. An artful blend of live-action and animation brings nine songs to life. There's "Belly Button," a clay animated clapper of nifty rhythms, and "Late Last Night," where kids are magically transported to wherever their footwear takes them. Most fun is "Skateboard," where a boy and his mom skate in highly improbable landscapes.

Joe Scruggs in Concert (ShadowPlay, 51 min., ages 3–10) Joe's second video shows the singer-songwriter in concert with his friends. And a good show they put on, too. "Deep in the Jungle" starts things off with a trio of teasing monkeys who get their just deserts. Other highlights include "Bahamas Pajamas," with Joe's mixed-up back-up birds, and "Goo Goo Ga Ga," starring a baby who can charm monsters. Throughout, Scruggs's family audience joins in by singing, clapping, and signing songs both traditional and original. As funny as he is musically talented, Scruggs gets his audience laughing a lot, too.

Kids' Guitar (Homespun, 90 min., ages 6–14) Award-winning instructor Marcy Marxer presents ten lessons in acoustic guitar aimed at young viewers. After a brief intro on choosing a guitar, its parts and tuning, Marxer teaches a two-chord song, "Skip to My Lou." Seven other songs are included so that by tape's end, the viewer should know six chords and three keys. The video is nicely staged and shot, with Marxer's approach suitably low-key and encouraging. *Kid's Guitar 2* includes "Shortenin' Bread" and seven other songs. See also *Making and Playing Homemade Instruments* in this section.

Kidsongs series (Warner-Reprise, 25 min. each, ages 3–8) This live-action sing-along series features classic children's songs, singably arranged and colorfully staged. Starring both kids and adults, each video is built around a theme (sports, animals) or visits a kid favorite (circus, camp). Here are some of the best:

Cars, Boats, Trains, and Planes Kids' fascination with vehicles is parlayed into fun music videos celebrating the likes of cars and trucks, balloons and skateboards. There's a good range of musical styles, too, from folk tunes ("Daylight Train") to rock ("I Got Wheels") to pop ("Up, Up and Away"). This video clicks with real kids in scenes that move along as surely as the vehicles themselves.

I'd Like to Teach the World to Sing A spin of the globe in the title song sends the kids off to different countries and their children's songs. "London Bridge," "Frere Jacques," and "Kumbaya" are some of the favorites. New tunes include "Sakura," set on a Japanese footbridge and "Los Pollitos," at a Mexican fountain. A nicely-paced collection of international music videos.

* **Very Silly Songs** In a visit to SillyDillyVille, the kids meet Willy and Jilly, who in turn introduce them to the "Purple People Eater," "Michael Finnegan," and "Jimalong Joe." All this alliteration is half the fun of the video, the other half being its rib-tickling tunes. From the introductory "The Name Game" to the finale dancer "The Silly Willy," this video gets preschoolers singing, hopping, clapping, and naturally laughing.

Let's Sing Along! (LCI, 25 min., ages 2–6) An invitation to imaginative play via original music. With no adult in sight, Brett and friends engage the viewer in sound-oriented make-believe.

One segment, for example, has Brett in extreme close-up as he sings about sneezes, yawns, and cheers. Another encourages the viewer to act like different animals, still another plays with rhythmic nonsense syllables. Like Brett's other videos, *Ring Along, Sing Along* and *You on Kazoo,* this production is well-done, interactive fun.

***Linda Arnold's World of Make Believe** (A&M, 48 min., ages 4–9) A magical mystery tour de force. This video showcases Arnold's talents in puppetry and theater arts as well as singing and songwriting. Before an audience of young children, she and her daughter sing of their favorite things ("Popcorn," "Make Believe," "Love") and characters ("The Merry Prankster," "Teddy Bear"). And there's a little mystery to solve, thanks to the Merry Prankster. On a set dusted with her trademark confetti, Arnold entrances her audience both there and at home.

Little People: Favorite Songs (R&G, 30 min., ages 2–6) A dozen classics are treated to animated visuals of the Little People family. The baby plays with blocks during "The Alphabet Song," and all the kids prepare for a visit from Grandma in "She'll Be Comin' Round the Mountain." Other good clips include "Animal Fair," "This Old Man," and "Pop Goes the Weasel."

Making and Playing Homemade Instruments (Homespun, 60 min., ages 6–14) Music performers/instructors Marcy Marxer and Cathy Fink show the makings of nine inventive instruments. Professionally demonstrated and shot, the video includes the likes of an oatmeal-box banjo, bottlecap castanets, and a washtub bass. Surprisingly musical, the instruments will often need adult assistance in construction. See Marxer's other tape in this section, *Kids' Guitar.*

Mother Goose Video Treasury, Volumes 1–4 (J2, 30 min., ages 3–6) Series of videos that bring nursery rhymes to life with "puppetronics." In *The Humpty Dumpty Collection*, this means Jack Be Nimble can jump over an eight-foot candle and Tom Thumb dance in Mother Goose's hand. Cheryl Rhoads makes a fine Mother Goose, introducing the likes of Old King Cole and

Peter Piper. Despite some unattractive puppets, the series is indeed a treasury of well-orchestrated songs and rhymes.

Muppet Video Series: The Kermit and Piggy Story (Walt Disney, 60 min., ages 3–10) One of several videos collecting the best of "The Muppet Show." Kermit and Miss Piggy's "love" affair is traced from their first gig together—performing "Temptation" with a barnyard chorus—to the finale "Ukulele Lady," a quintessential "Muppet Show" number with Miss Piggy on a tropical moon. In between are assorted sketches, quips, and songs with such celebrity guests as Raquel Welch and Tony Randall. Miss Piggy steals the show, however, with her Carmen Miranda routine. Musical Muppet mania at its best.

Nonsense and Lullabyes: Nursery Rhymes (FHE, 27 min., ages 2–5) Animator Michael Sporn takes these two staples of childhood and treats them to fresh interpretations, with original music by Caleb Sampson. Best of the "nonsense" includes "The House That Jack Built" told as an animated slide show, "Hey Diddle Diddle's" charming artwork, and "The Chivalrous Shark," a true man-eater who eschews women and children. And of the lullabyes, most entrancing are "Twinkle Twinkle Little Star," with new music and lyrics, and "Winken Blynken and Nod," which is beautifully drawn and animated, and then read by Linda Hunt. Also available is *Nonsense and Lullabyes: Poems For Children.*

Orchestra, The (Silo-Alcazar, 40 min., ages 8–14) An award-winning introduction to classical music and its instruments. Accompanied by drawings and music, Peter Ustinov ably narrates this sound, in every sense, education. Opening with a section on the feelings and range expressed in music, the video goes on to explore the orchestral components of strings, woodwinds, brasses, percussion, and the conductor.

Pete Seeger's Family Concert (Sony Kids Video, 45 min., ages 4–8 with family) From the banks of his beloved Hudson River, Seeger performs over a dozen songs to an audience sprawled on blankets and playing on the beach. As billed, it's a family audience that, smiling and singing along, seems as much a part of the concert as the venerable folk singer-songwriter. Blending kids' favorites ("Skip to My Lou," "She'll Be Comin' Round the Mountain"), folk classics ("This Land Is Your Land"), and original river tunes from his clean-up campaign, Seeger adds a little storytelling to the mix. Simply clad and set in a sun-dappled cluster of trees, the gentle persuader looks every bit the old-time musician he speaks of so nostalgically in the video. Simply heartwarming.

*__Raffi: A Young Children's Concert__ (A&M, 45 min., ages 3–7) First video of the best-selling children's artist sings with musical delights. "Down by the Bay" and "Baby Beluga" are here, as well as "Six Little Ducks," the silly "In the Store," and active "Wiggle My Wiggles Away." Raffi's easygoing style and encouragement to sing, clap, hum, tweet, and stomp along build rapport with his family audience. Since this comes through even on video, Raffi should continue to garner fans despite his crossover to adult music.

*__Raffi in Concert With the Rise and Shine Band__ (A&M, 50 min., ages 3–7) Again, Raffi's winning combination of warmth and silliness shines in this concert. From the beautiful "Rise and Shine" to the nonsense of "Apple and Bananas" to the poignant "Baby Beluga," his repertoire reflects the full humanity of children. Raffi appreciates kids' rhythms and relates to them in all their moods. And the video is very well produced. Artistic attention is paid to every detail—from cover art to set design, from the music's creation to its staging. Overall, a warm and fun experience.

*__Rappin' n' Rhymin'__ (Hanna-Barbera, 30 min., ages 3–8) "Y' like rap? Then follow me!" is the rousing opener to this video encouraging kids to get busy. Two teenagers lead younger children in high-energy dances to seven rap songs. The sets are colorful and well designed, and the original music is as fresh as the kids. Between music videos are counting and rhyming scenes with Yogi Bear and Fred Flintstone. Both educational and active, *Rappin'* offers something unique in kidvid.

Ring Along, Sing Along! (LCI, 30 min., ages 3–6) From the creators of *Let's Sing Along* and *You on Kazoo* comes this ring-a-ding video that includes a jingle bell toy. Once again kids themselves star, and shine they do. Like a bright-eyed Peter Pan, Brett leads them in song and dance and game in seven imaginative segments. The title song is a fun rocker where the kids introduce themselves. Then it's a game of skipping, jumping, singing, and ringing in "You and Me." And in "Ring Those Bells," they follow the leader shaking fast, funny, and cra-a-a-a-zy rhythms. Beautifully produced and set in meadows and comfortable classrooms, *Ring Along* is an invitation to musical play. Because the toy has small parts, however, you should supervise its use by young children.

Rory Story, The (Sony Kids Video, 60 min., ages 4–9) Playing like an extended episode of "The Muppet Show," this video has children's recording artist Rory interacting with an array of showbiz puppets. The story chronicles her rise from member of The Incredible Piglets to solo singer. While the story scenes lag a bit in pace and humor, Rory's original tunes sparkle . . . as does her engaging, childlike voice. Best cuts are "Toys," a breezy soft-shoe, the pop lullaby "Time for Bed," and "You're Drivin' Me Crazy," sung from both parent and child points of view.

Sesame Songs series (Random House, 30 min., ages 2–6) Knowing music to be one of Sesame Street's strengths, Random House Home Video created this special line. Each video includes a lyrics sheet and collects music videos for little ones, pulled together by a story or game. Here are some of the best examples . . .

* **Dance Along** Fun and funny dances presented by Big Bird in a Club MTV-type setting. Your child can follow-the-leader, act the bat, freeze dance, and do the Pigeon with Bert. There's even an Any Way You Feel dance with rhythms from around the world.

Rock ’n Roll Deejay Jackman Wolf is taking requests and spinning Sesame Street’s best boppin’ tunes. Chrissy and the Alphabeats sing out “You’re Alive!,” while The Talking Hands show all the communicating we can do with our hands. And who but Bert would request, and dedicate to his paperclip collection, “It’s Hip to Be Square?” Big beat fun fun fun till your daddy takes the TV away.

Sing, Hoot, and Howl As the title hints, the theme is animals. And the music is great, with a dozen songs deftly blending education and entertainment. Included are classics, such as the hard-working “Cow Dog” and “The Chicken and the Egg,” and other fun numbers, such as “Pig’s Love Song” and “Baa Baa Bamba.”

* **Sing Yourself Silly** Great collection of Sesame Street classics. “The Ladybug Picnic” animates the little buggers jumping rope and having sack races. “The Everything in the Wrong Place Ball” stars Oscar and date comically misdressed for a dance. And the video wraps up with “Put Down the Duckie,” sound advice from the likes of Paul Simon, Pete Seeger, Danny De Vito, and Rhea Perlman to Ernie, whose sax has a mysterious squeak. Great fun from start to finish.

Simply Mad About the Mouse (Walt Disney, 35 min., ages 3–10) To dust off their classic movie tunes, Disney asked the likes of Billy Joel and Harry Connick, Jr., to make new, hi-tech music videos. Of the eight included on the tape, top cuts are The Gipsy Kings’ “I’ve Got No Strings,” with animation as stunning as the guitar work, Soul II Soul’s “Kiss the Girl,” full of tropical colors and sounds, and L.L. Cool J’s smoothly rapped “Who’s Afraid of the Big Bad Wolf?” filmed in stylized black and white. A family treat.

Sing-Along, Dance-Along, Do-Along (Lorimar, 30 min., ages 3–6) Jim Henson Play-Along Video that gets kids up and moving to ten traditional and new tunes. The “Gob-a-goo Rap” encourages clapping, stomping, bowing, etc., to prevent becoming a gob-a-goo. A variation on “In a Cabin in the Woods” places a monster in the window. And calypso reigns in the “All Pig Island

Band." A fitting finale is "You Gotta Sing""when the spirit says sing," not to mention hop, sneeze, swim, and hug. Bright, interactive fun.

Sing-Along Songs series (Walt Disney, 30 min. each, ages 3–10) This wonderful line capitalizes on a Disney strength—wedding animation to original music. Both are of such quality that their union seems natural and nearly perfect. Plus these music videos for kids often have lyrics printed on screen. So they're more than fun, they're reading-enhancing. The best, each including about ten songs, are . . .

Bare Necessities, The Title cut, of course, is the most memorable tune from *The Jungle Book.* Other hits include the "Cinderella Work Song," where the mice call her Cinderelly, Winnie the Pooh tune about the "willy nilly silly old bear," and "Ev'rybody Wants to Be a Cat," Dixieland jazz from *The Aristocats.* Wrapping up the tape is Burl Ives singing "The Ugly Bug Ball" from *Summer Magic,* which cuts to great vintage animation from Max Fleischer.

Be Our Guest Title tune from *Beauty and the Beast* gives a nod to Busby Berkeley as everything from dishes to flowers, napkins and flatware dance in excitement to entertain Belle. Other highlights include *Mary Poppins's* "Spoonful of Sugar" and "Chim Chim Cheree," "Once Upon a Dream" from *Sleeping Beauty,* and the simply lovely, Oscar-winning song, "Beauty and the Beast." (Check with your dealer on this title's availability.)

Heigh-Ho Title clip from *Snow White and the Seven Dwarfs* includes the classic shot of Dopey with faceted-diamond eyes. Next up is Winnie the Pooh's stoutness exercises in "Up Down and Touch the Ground." Best of the remaining eight songs are "The Dwarf's Yodel Song," Donald Duck's "The Three Caballeros," and Peggy Lee's "The Siamese Cat Song" from *Lady and the Tramp.*

Under the Sea The best of nine songs are culled from *The Little Mermaid* ("Under the Sea," "Kiss the Girl"), *Peter Pan* ("Never Smile at a Crocodile"), and vintage Disney cartoons.

"At the Codfish Ball," for example, shows mer-kids peeking in on a sea life dance where "water beetles twist and shout."

You Can Fly! Who can resist the joy of this song-scene from *Peter Pan?* It gets the movie *and* this collection of Disney music videos off to a great start. The finish is fun, too, as the crows from *Dumbo* sing "When I See an Elephant Fly." In between are musical clips from *Pinocchio* ("I've Got No Strings"), *Winnie the Pooh* ("Little Black Rain Cloud"), and *Lady and the Tramp.*

* **Zip-A-Dee-Doo-Dah** A real feel-good tape. Title cut from *Song of the South* is, of course, quintessentially upbeat. Other fun clips include *Alice in Wonderland*'s "The Unbirthday Song" and "Bibbidi-Bobbidi-Boo" from *Cinderella.* Finale number is *Snow White and the Seven Dwarfs'* "Whistle While You Work." An animated music video treasury.

Tickle Tune Typhoon: Let's Be Friends (Celebrity, 50 min., ages 6–10) Concert of the award-winning Seattle band whose strong musicianship and songwriting have production values to match in this first video. Their typhoon tunes do more than tickle, they engage the mind and spirit in themes ranging from race and prejudice to the environment and differently-abled. And they draw on a variety of styles to create music for families, especially those with kids too old for preschool and too young for pop.

***Toddler Treasury, A** (Creative Learning Products, 20 min., ages 2–5) Created by Nickelodeon founder Dr. Vivian Horner, *Treasury* offers a wealth of nursery rhymes set to simple animation. Both children and adults sing the likes of "Itsy Bitsy Spider" and "Twinkle Twinkle Little Star." Visuals for the sixteen rhymes, songs, and games have nice touches, too, like jogging suits for the "Three Blind Mice." Attractive and well-paced, the video is very toddler-friendly.

Wee Sing series (Price/Stern/Sloan, 60 min. each, ages 3–8) Built on the best-selling staples of children's music, *Wee Sing* books and audiocassettes, the videos continue a tradition of excellence. Their arrangements are solid and singing style so unpretentious, anyone can join in the fun. Created by partner moms, these are the finest Wee Sing videos . . .

* **Big Rock Candy Mountains** A young girl visits the mountains where over twenty songs are woven into the story of "Little Bunny Foo Foo." Also integrated are lessons in nutrition and behavior. And there are plenty of jokes and riddles to break things up. With songs like "Fiddle-Dee-Dee" and "Jimmy Crack Corn," the overall effect is one of a fun party.

In Sillyville The colors are at odds in the land of Yellow Spurtlegurgles and Green Jingleheimers. So a boy and girl team up with peacemaker Sillywhim to bring the colors together again. Along the way are staged twenty silly songs, such as "Boom Boom, Ain't It Great to Be Crazy?" and "Do Your Ears Hang Low?" As bright as a new crayon box, this video has songs to match.

* **King Cole's Party** Over twenty familiar nursery rhymes come to life in this exceptional music video. The occasion is King Cole's Party and invited are the likes of Humpty Dumpty, Peter Piper, and Jack and Jill. The effect is a musical variety show for little ones, very well produced and brimming with song, dance, and comedy.

Wee Sing: The Best Christmas Ever—see Holidays: Christmas

***You on Kazoo!** (LCI, 30 min., ages 3–8) With included kazoo, this video uses the humble hummer to bring out your child's imagination. Seven children—no adult in sight—invite the viewer to play, sing, and dance along. "You can do it," exhorts one song, "nothing to it, just kazoo it!" The kids' pretend play includes animals, airplanes, and adventures in the Land of Make-Believe. As interactive as videos get. See also *Let's Sing Along* and *Ring Along, Sing Along!*

APPENDIX A

THE BEST CONTINUING SERIES

What to look for in kidvid? The short answer would be these names on the label. Their very best are reviewed here, but you can't go wrong with any of their videos. Keep them in mind when looking for new videos for your children.

The Baby-Sitters Club. The Ann B. Martin books for 6–to–12–year–old girls are faithfully adapted in a live-action format. The current titles are reviewed here in the Book-Based section under *Baby-Sitters Club* and in Holidays: Christmas section is *The Baby-Sitters Club Special Christmas.*

Children's Circle. When producer Mort Schindel founded Weston Woods in the early 50s, he was virtually the only one adapting children's literature. Now that his films are being released on home video under the Children's Circle banner, his mission has even more outreach. "We try to promote literacy by leading children back to good books," he says, "and to encourage them not just to learn how to read but to become readers." His mission is accomplished by faithfully adapting the best picture books from around the world. Reviewed in this book are: *What's Under My Bed?* (Holidays: Halloween); *The Emperor's New Clothes* (Folk and Fairy Tales); and in the Book-Based section are *Corduroy, Doctor De Soto, Five Stories for the Very Young, Happy Birthday Moon, Joey Runs Away, Madeline's Rescue, The Maurice Sendak Library, Norman the Doorman, Owl Moon, The Pigs' Wedding, The Robert McCloskey Library, The Snowman,* and *Stories From the Black Tradition.*

Rabbit Ears Productions. Like King Midas, the star of one of his videos, producer Mark Sottnick has a golden touch. His talent is selecting just the right talents in adaptation, illustration, narration, and music to bring a story to life. His first series, Storybook Classics, runs the gamut from *Just So Stories* to Hans Christian Anderson fairy tales to American tall tales. Both known and new stories from around the world are collected in the We All Have Tales line. And most recently is American Heroes & Legends with the likes of John Henry and Sacajawea. Reviewed here are these Storybook Classics in the Book-Based section: *The Elephant's Child, How the Leopard Got His Spots,* and *The Velveteen Rabbit;* in Folk and Fairy Tales: *Br'er Rabbit and the Wonderful Tar Baby, The Emperor and the Nightingale, The Emperor's New Clothes, Paul Bunyan, Pecos Bill, Thumbelina,* and *The Ugly Duckling;* and in Holidays: Halloween is *The Legend of Sleepy Hollow* and in Christmas are *The Gingham Dog and the Calico Cat* and *The Tailor of Gloucester.* These We All Have Tales videos are in the Folk and Fairy Tales section: *Anansi, The Boy Who Drew Cats, Finn McCoul, The Fool and the Flying Ship, King Midas and the Golden Touch, Koi and the Kola Nuts, Peachboy* and *Puss in Boots. Annie Oakley* and *Stormalong* from American Heroes & Legends, are also in Folk and Fairy Tales.

Sesame Street. It was a Sunny Day indeed when Children's Television Workshop swept away the clouds of boring educational TV. Sesame Street paved the way for a revolution in children's programming, one that showed television could be good for kids . . . and that education could be highly entertaining. Sesame Street videos collect the best segments and arrange them along such themes as letters, games, and self-esteem. The best of these collections are reviewed here primarily in the Educational section under *Sesame Street*, plus under *Sesame Songs* in Music, and under *Sesame Street Special* in Holidays: Christmas.

Shelley Duvall. Few producers have shown the dedication and talent that Shelley Duvall applies to children's video. She has a knack for selecting the best people for the best projects. And it all stems from her lifelong love of storybooks. You'll find her Faerie Tale Theatre and Shelley Duvall's Tall Tales and Legends videos in the Folk and Fairy Tale section. And her most recent series, Shelley Duvall's Bedtime Stories, is covered in the Book-Based section.

Sing-Alongs. Too numerous to mention individually are a number of producers of quality music videos for children. See the Best Sing-Along Videos in Appendix B for their titles.

Smarty Pants. In 1941, Norman McLaren opened one of the world's most respected and innovative animation studios, that of The National Film Board of Canada. Since then, animators have been drawn there to create not only traditional cel animation but new techniques with everything from beads to sand to computers. Now these treasures are being released on the Smarty Pants label, most videos containing several stories. The best collections, reviewed here in the Animation section, are *The Cat Came Back, The Dingles, Every Child, Peep and the Big Wide World, Sea Dream,* and *The Tender Tale of Cinderella Penguin.*

Stories to Remember. Beautifully-crafted animation is the hallmark of this Lightyear Entertainment line. And what looks like an eclectic collection of tales, according to producer Joshua Greene, is really "a carefully chosen family of great stories from around the world, stories which touch the heart of young and old." Based on the best in children's literature, Stories to Remember has attracted the best in music composition and narration as well—the likes of Ernest Troost, Mia Farrow, Kevin Kline, and Sigourney Weaver. Reviewed here are *Baby's Bedtime, Baby's Morningtime, Baby's Storytime,* and *Merlin and the Dragons* in the Book-Based section; *Beauty and the Beast* and *The Snow Queen* in Folk and Fairy Tales; and *Pegasus* in Animation. Coming up are *The Wild Swans,* a Hans Christian Andersen tale, and *People,* based on the wonderful Peter Spier book.

Timeless Tales From Hallmark. Animated by Hanna-Barbera, these videos are simply good stories well told. They're classic fairy tales treated to new songs and, in some cases, a new twist. Naturalistically drawn and written, they make a fine introduction to the genre for the very young. Reviewed here in the Folk and Fairy Tales section are: *The Elves and the Shoemaker, Puss in Boots, Rumpelstiltzkin,* and *The Steadfast Tin Soldier.*

Walt Disney Cartoon Classics. As deserved as all the attention on Disney's animated features is, it tends to obscure their treasure chest of seven-minute classics. Collected on video as "Cartoon Classics" they're presented as star vehicles for Disney's stable of characters. And they showcase animation so inventive and entertaining, it's been a favorite of children and adults for generations. Reviewed here in the Animation section are: *Fun on the Job, Here's Donald, Here's Mickey, Nuts About Chip 'n Dale, Silly Symphonies, Starring Chip 'n Dale, Starring Mickey and Minnie,* and the Limited Gold Editions (still available as rentals) *Donald's Bee Pictures* and *Life With Mickey;* and in Holidays: Halloween are *Donald's Scary Tales* and *Halloween Haunts.*

Walt Disney Mini-Classics. Disney also has an impressive collection of animated featurettes, released on video as "Mini-Classics." Reviewed in the Book-Based section are: *The Prince and the Pauper* and *Winnie the Pooh* shorts; *Mickey's Christmas Carol* is in Holidays: Christmas; and *Mickey and the Beanstalk* is in Folk and Fairy Tales. You'll find the remainder reviewed in Animation: *Ben and Me, Peter and the Wolf,* and *The Reluctant Dragon.*

WonderWorks. This award-winning PBS series is known for excellence in family drama. And while that may sound stuffy, what it means is that you *and* your kids will become engrossed in the best of its stories. Reviewed here, in the Family Topics—Dramas and Documentaries section are: *African Journey, Almost Partners, And the Children Shall Lead, Bridge to Terabithia, Brother Future, Gryphon, Jacob Have I Loved, A Little Princess, Maricela, The Mighty Pawns, Runaway, Sweet 15,* and *You Must Remember This.*

APPENDIX B

A BAKER'S DOZEN OF BEST LISTS

Children's video, like literature, fills many functions in entertainment, education, and guidance. The following lists suggest the best dozen videos in categories that reflect these uses and special needs. And many of them list videos that can help make television viewing a positive experience for your child.

The Best First Videos for Toddlers

Probably by the age of two, a child is ready to be introduced to television. The following are home videos offering age-appropriate information, art, music, and pace for the very young.

Are You My Mother?...
Baby Songs videos, especially Volume 2
Baby Vision
Baby's Bedtime/Baby's Morningtime
Barney and the Backyard Gang videos
Five Lionni Classics
Good Morning Sunshine
Nonsense and Lullabyes
Peek-a-Boo
Sesame Songs and Sesame Street videos
The Tales of Beatrix Potter
A Toddler Treasury

The Best Bedtime Videos

Like a bedtime book, a video at night should be engaging but not too stimulating. The following are videos either made specifically for the purpose or relaxing in nature to help children settle down.

Baby's Bedtime
Foxy Fables
The Land of Pleasant Dreams
Maple Town
Paddington Bear
Sesame Street: Bedtime Stories and Songs
Shelley Duvall's Bedtime Stories
The Tailor of Gloucester
The Tales of Beatrix Potter
Thomas the Tank Engine and Friends
The Velveteen Rabbit
Winnie the Pooh videos

The Best Birthday Party Videos

Showing a video at a birthday party gives everyone a break. Animation and sing-along tapes are favorites, but you might also consider the following, which either feature a party or have a festive atmosphere.

Barney's Birthday
The Berenstain Bears and Too Much Birthday
Charles the Clown
Circus ABC's/Circus 123's
Clifford's Fun With Numbers
Draw Squad: Moonbot's Birthday Party
It Zwibble: Earth Day Birthday
The Maurice Sendak Library
Preschool Power videos

The Railway Dragon
Sesame Street: Big Bird's Favorite Party Games
Wee Sing: King Cole's Party

The Best Dinosaur Videos

Ever a favorite with kids, dinosaurs have found their way to many a videotape. The following either star or educate on the prehistoric beasts.

Barney and the Backyard Gang videos
Denver the Last Dinosaur
Dinosaur!
Dinosaurs!
Fantasia
It Zwibble: Earth Day Birthday
The Land Before Time
Mister Rogers Talks About Dinosaurs . . .
Newton's Apple, Volume 1
Reading Rainbow: Digging Up Dinosaurs
Son of Dinosaurs
The World's Greatest Dinosaur Video

The Best Detective Story Videos

Like good mystery books, these videos help develop young minds as they follow clues, piece them together, and solve puzzles. All are good detective stories that reward paying attention and using your intellect.

Almost Partners
Baby-Sitters Club: Claudia and the Missing Jewels
Berenstain Bears: Missing Dinosaur Bone
Clifford's Fun With Shapes and Colors
Dirkham Detective Agency/Mystery at Fire Island
Edgar Allan Poe's The Gold Bug

Encyclopedia Brown (2)
Fun in a Box: Ben's Dream
Nancy Drew
Peanuts: It's a Mystery, Charlie Brown
Scooby and Scrappy-Doo
Sherlock Holmes: The Priory School

The Best Sing-Along Videos

Everybody likes to sing. So you may find yourself drawn to these videos designed for children. They have attracted family audiences with excellence—in song selection, arrangement, and presentation.

Baby Songs videos
Barney and the Backyard Gang videos
Bill Harley: Who Made This Mess?
Ella Jenkins Live!
Joe's First Video
Kidsongs
Linda Arnold's World of Make Believe
Raffi (2)
Sesame Songs
Sing-Along Songs
Wee Sing videos
You on Kazoo!

The Best Videos With Positive Female Roles

Television fails women by stereotyping and under representation–for every female on TV there are three males. These videos starring girls and women present them as strong, independent, intelligent, and active. Here are the best videos for children that offer positive female role models.

Anne of Green Gables
Annie Oakley + Darlin' Clementine
Baby-Sitters Club, especially *Kristy and the Great Campaign*
Beauty and the Beast
Berenstain Bears: No Girls Allowed
The Dancing Princesses
The *Dot* series
Free to Be . . . You and Me
Frog Girl
The Snow Queen
Sweet 15 + Maricela
A Woman's Place

The Best Videos With Positive Black Roles

Like women, blacks are underrepresented and stereotyped on television. The following videos for children can help counter this media failing.

And the Children Shall Lead
Brother Future
The House of Dies Drear
Koi and the Kola Nuts
The Mighty Pawns
The Red Shoes
Roots
Runaway
The Speeches of Martin Luther King, Jr.
Stories From the Black Tradition
Where in the World: Kids Explore Kenya
You Must Remember This

The Best Videos for the First Day of School

Some of the following directly address the anxieties of the first day of school. Others are good examples of what your child can expect in a typical school day.

Baby Songs 1
Baby Songs 8
Barney Goes to School
Bill Cosby's Picture Pages
Imagine That!
Mister Rogers: When Parents are Away
Preschool Power videos
Richard Scarry's Best ABC Video Ever
Sesame Street: Big Bird's Favorite Party Games
Sesame Street: Getting Ready for School
Who Will Be My Friend?
Workout With Daddy and Me (+ Mommy and Me)

The Best Videos Fostering Imagination

Like a good book, any good video sparks a child's imagination. The following most directly celebrate or encourage imaginative activity.

Barney and the Backyard Gang videos
Draw Squad + other Mark Kistler videos
Dr. Seuss: One Fish, Two Fish, Red Fish, Blue Fish
Fun in a Box videos
Imagine That!
Let's Sing Along
Mister Rogers videos
My Fun Pack videos
Peter Pan (2)
Scholastic Blue Ribbon Storybook Videos
Sea Dream . . .
Squiggles, Dots and Lines

The Best Videos Showing Cooperation

Children who see people cooperate on television tend to be more cooperative themselves.† The following videos show children–and animals and trains!–learning to get along and work together for a common goal.

Baby-Sitters Club videos
Elizabeth and Larry/Bill and Pete
50 Simple Things Kids Can Do . . .
Fraggle Rock
The Gingham Dog and the Calico Cat
The Lollipop Dragon
Mine and Yours
Preschool Power 3
Scamper the Penguin
The Story of 15 Boys
Thomas the Tank Engine and Friends
The Wizard of Oz

The Best Videos Encouraging Nonviolence

To counter the overwhelming incidence of violence on television, these videos suggest nonviolent means of problem-solving. In some, aggression is possible but averted; in others characters work out problems nonviolently. And in *The Butter Battle Book,* a cold war parable, the threat of violence is shown to have its own destructiveness.

Babar the Little Elephant
The Berenstain Bears Get in a Fight
The Butter Battle Book

†Berk, Laura E. *Child Development.* Needham Heights, Mass.: Allyn and Bacon, 1989.

Finn McCoul
Fraggle Rock
Madeline and the Bad Hat
The Mighty Pawns
Mister Rogers: What About Love?
The Reluctant Dragon
Stories From the Black Tradition
SuperTed: Leave It to Space Beavers
Thomas the Tank Engine and Friends

The Best Videos With Creative Problem-Solving

In addition to the detective story videos listed earlier, these offer characters who face problems with clear and clever thought.

Abel's Island
Anansi
The Astronomers
Berenstain Bears videos
Bill Cosby's Picture Pages
Br'er Rabbit and the Wonderful Tar Baby
Doctor De Soto . . .
The Fool and the Flying Ship
How the Leopard Got His Spots
Mike Mulligan and His Steam Shovel
Peep and the Big Wide World
Winnie the Pooh and the Honey Tree

APPENDIX C

CHILDREN'S VIDEO SUPPLIERS

Many of the videos in this book are readily available in retail outlets—discount, department, toy, music, book, and video stores. If you cannot find one on the shelves, check with your video dealer to order it. If *that* fails, you can contact the supplier directly using this list of addresses and telephone numbers. Some are set up to take your order, others will direct you to the nearest retailer.

A&M Video 800-888-5301
1416 North LaBrea
Hollywood, Calif. 90028

Bridgestone Group 800-523-0988
1979 Palomar Oaks Way
Carlsbad, Calif. 92009

Butterfly Video 800-433-2623
P.O. Box 184
Antrim, N.H. 03440

CBS-Fox Video 800-800-2369
1211 Avenue of the Americas
New York, N.Y. 10036

Celebrity Home Entertainment 800-445-8210
P.O. Box 4112
Woodland Hills, Calif. 91365-4112

Children's Circle 800-543-7843
389 Newtown Turnpike
Weston, Conn. 06883
(also distributed through Wood Knapp)

The Children's Group 800-668-0242
561 Bloor Street West #300 (The Children's Book Store
Toronto, Ontario Distribution, N.Y.)
CANADA M5S 1Y6

Children's Television Workshop (CTW) 800-822-1105
Video Services
50 Leyland Drive
Leonia, N.J. 17605

Columbia TriStar 818-953-7900
P.O. Box 310
San Fernando, Calif. 91341

Concept Associates 800-333-8252
7910 Woodmont Avenue #1214
Bethesda, Md. 20814

ConsumerVision 800-756-8792
149 Fifth Avenue 8th floor
New York, N.Y. 10010

Creative Learning Products 800-262-2437
3567 Kennedy Road
South Plainfield, N.J. 07080

Crown Video—see ConsumerVision

David Eagle Productions 213-286-7002
1551 South Robertson Boulevard #203
Los Angeles, Calif. 90035

Encounter Video 800-677-7607
2580 NW Upshur #202
Portland, Ore. 97210

ESPN Home Video 203-585-2000
ESPN Plaza
Bristol, Conn. 06010-7454

FHE/LIVE 800-752-9343
c/o Comar Distributors
2030 East University Drive
Rancho Dominguez, Calif. 90220

Fries Home Video 213-466-2266
6922 Hollywood Boulevard
Los Angeles, Calif. 90028

Fusion Video 800-338-7710
17311 Fusion Way (Orders only)
Country Club Hills, Ill. 60478 708-799-2214
(Information on locating many videos)

Golden Book Video 800-236-7123
1220 Mound Avenue
Racine, Wis. 53404

Golden Glow Recordings 800-394-0493
P.O. Box 601
Kensington, M.D. 20895

GoodTimes Home Video 212-951-3100
16 East 40th Street
New York, N.Y. 10016

Hanna-Barbera Home Video 213-969-1211
3400 Cahuenga Boulevard
Hollywood, Calif. 90068-1376
(Some videos with Turner Home Entertainment)

Homespun Video 800-338-2737
P.O. Box 694
Woodstock, N.Y. 12498

J2 Communications 213-474-5252
10850 Wilshire Boulevard #1000
Los Angeles, Calif. 90024

JCI Video 818-593-3600
21550 Oxnard Street #920
Woodland Hills, Calif. 91367

Kidvidz 617-243-7611
618 Centre Street
Newton, Mass. 02158

Kultur Video 800-458-5887
121 Highway 36
West Long Branch, N.J. 07764

LandyVision 800-777-9755
11 Hill 99
Woodstock, N.Y. 12498

LCI 800-284-3948
578 Post Road East #520
Westport, Conn. 06880

Lee Mendelson Film Productions 415-342-8284
1440 Chapin Avenue
Burlingame, Calif. 94010

Lightyear Entertainment 800-229-7867
350 Fifth Avenue #5101
New York, N.Y. 10118

Lorimar Home Video—Some are now with Warner

Lyons Group 800-527-4747
P.O. Box 8000
Allen, Tex. 75002-1306

Made-to-Order Productions 312-525-7701
636 Deming Place
Chicago, Ill. 60614

Mark Kistler Productions 800-321-3729
P.O. Box 361
Oceanside, Calif. 92054

MCA Home Video 818-777-5539
70 Universal City Plaza
Universal City, Calif. 91608

Media Home Entertainment—see Video Treasures

MGM/UA Home Video 213-280-6000
10000 West Washington Boulevard
Culver City, Calif. 90232

Mid-Com, Inc. 812-473-0533
555 Byerson Drive
Newburgh, Ind. 47630

Miramar Producations 800-245-6472
200 Second Avenue West
Seattle, Wash. 98119

Morris Video 800-843-3606
2730 Monterey #105
Torrance, Calif. 90503

MPI Home Video 800-323-0442
15825 Rob Roy Drive
Oak Forest, Ill. 60452

New World Video 310-444-8100
1440 South Sepulveda
Los Angeles, Calif. 90025

Pacific Arts 800-333-0041
11858 LaGrange Avenue
Los Angeles, Calif. 90025

Paramount Home Video 213-956-5000
5555 Melrose Avenue
Hollywood, Calif. 90038

Picture Pages 800-336-2360
2313 East Carson
Pittsburgh, Pa. 15203

Price/Stern/Sloan 800-421-0892
11150 Olympic Boulevard #650
Los Angeles, Calif. 90064

Prism Entertainment 310-277-3270
1888 Century Park East #350
Los Angeles, Calif. 90067

Public Media Video 800-262-8600
5547 North Ravenswood Avenue
Chicago, Ill. 60640-1199

R&G Video 818-907-3888
14724 Ventura Boulevard #200
Sherman Oaks, Calif. 91403

Rabbit Ears Productions 203-857-3760
131 Roywayton Avenue 800-243-4504
Rowayton, Conn. 06853 (Listening Library)
(Videos distributed through Uni and Columbia TriStar)

Random House Home Video 800-733-3000
400 Hahn Road
Westminster, Md. 21157

Republic Pictures/Blackhawk Films 800-826-2295
12636 Beatrice Street (Orders)
Los Angeles, Calif. 90066 800-544-9852
(Customer Service)

Scholastic Video 212-505-3000
730 Broadway
New York, N.Y. 10003

Sea Studios 408-649-5152
810 Cannery Row
Monterey, Calif. 93940
(Available through the Nature Company) 800-227-1114

ShadowPlay Video 800-274-8804
P.O. Box 180476
Austin, Tex. 78718

Silo-Alcazar 800-541-9904
P.O. Box 429
Waterbury, Vt. 05676

Smarty Pants Video 216-221-5300
15104 Detroit Avenue #2
Lakewood, Ohio 44107

Smithsonian Folkways 800-443-4727
c/o Rounder Records
One Camp Street
Cambridge, Mass. 02140

Sony Kids Video 212-445-4321
P.O. Box 4450
New York, N.Y. 10101

Stabur Home Video 800-346-8940
11904 Farmington Road
Livonia, Mich. 48150

Starmaker Entertainment 800-233-3738
151 Industrial Way East (Retailers only)
Eatontown, N.J. 07724 201-389-1020

Start Right Video 408-429-6452
10700 Santa Monica Boulevard #4-303
Los Angeles, Calif. 90025

Strand Home Video 800-468-7246
3350 Ocean Park Boulevard #205
Santa Monica, Calif. 90405

Summit Media Company 800-777-8668
27811 Avenue Hopkins Unit 1
Valencia, Calif. 91355

SVS—see Columbia TriStar

3E Communications 408-379-0589
15732 Los Gatos Boulevard #345
Los Gatos, Calif. 95032

Turner Home Entertainment (THE) 404-827-1013
P.O. Box 105366
Atlanta, Ga. 30348-5366

Vestron—See FHE/LIVE

VidAmerica 800-843-1994
60 Madison Avenue 12th floor
New York, N.Y. 10010

Video Treasures/Handleman 800-786-8777
500 Kirts Boulevard (Retailers only)
Troy, Mich. 48084

V.I.E.W. Video 800-843-9843
34 East 23rd Street
New York, N.Y. 10010

Walt Disney Home Video 800-227-9483
500 South Buena Vista
Burbank, Calif. 91521

Warner Home Video 818-954-6000
4000 Warner Boulevard
Burbank, Calif. 91522

Warner-Reprise Video 818-843-6311
3300 Warner Boulevard
Burbank, Calif. 91505

Wood Knapp 213-965-3500
5900 Wilshire Boulevard
Los Angeles, Calif. 90036

TITLE INDEX

ABOUT THE AUTHOR

Catherine Cella has covered children's video virtually from its inception. She reviews children's audio and video for *Billboard* magazine, is a contributing editor at *Parents' Guide to Children's Entertainment*, and a longtime reviewer at *Kids Today.* Formerly called Pennywhistle Press, the latter is a "USA Today for kids" published by Gannett for its newspapers nationwide.

As an advocate for both children and consumers, Cella co-authored the 1985 Consumer Guide book *Children's Video Tapes and Discs*, in addition to writing for their magazine *Video Times.* She has also reviewed kidvid in a syndicated column for parenting publications, for Gannett News Service, and in *Children's Video, Home Viewer,* and *Video Choice* magazines. She has published articles on children's video in such publications as *Video, Ms.*, and *Mothering.* And she has written on parenting topics for Johnson + Johnson, Mothering, and several parenting publications.

As a mother, former early childhood educator, and current classroom volunteer, Cella has always had a special place in her heart for children. She works for UNICEF and is a member of EDPRESS, a professional organization dedicated to education through print communication. She believes children's videos deserve critical review, using high standards of quality and bearing in mind the special needs of the age group. Respecting their opinions, she shares videos with children to gauge their responses.

Cella graduated from Macalester College with a degree in philosophy and minor in art history. She lives in Tennessee with her husband and son, where she enjoys reading, photography, international folk dancing, and oh yes, television.